Practical Magnetotellurics

The magnetotelluric (MT) method is a technique for probing the electrical conductivity structure of the Earth to depths of up to 600 km. Although less well known than seismology, MT is increasingly used both in applied geophysics and in basic research. This is the first book on the subject to go into detail on practical aspects of applying the MT technique.

Beginning with the basic principles of electromagnetic induction in the Earth, this introduction to magnetotellurics aims to guide students and researchers in geophysics and other areas of Earth science through the practical aspects of the MT method: from planning a field campaign, through data processing and modelling, to tectonic and geodynamic interpretation. The book contains an extensive, up-to-date reference list, which will be of use to both newcomers to MT and more experienced practitioners of the method. MT is presented as a young and vibrant area of geophysical research with an exciting potential that is yet to be fully realised.

The book will be of use to graduate-level students and researchers who are embarking on a research project involving MT, to lecturers preparing courses on MT and to geoscientists involved in multidisciplinary research projects who wish to incorporate MT results into their interpretations.

FIONA SIMPSON has held the position of assistant professor of experimental geophysics and geodynamics at Göttingen University where she teaches courses on pure and applied geophysics, and rheology of the Earth, since 2002. Dr Simpson has conducted MT fieldwork in Africa, Australia, Europe and New Zealand, and her research interests in addition to MT include deformation processes in the lithosphere and in the sub-lithospheric mantle.

KARSTEN BAHR has been a professor of geophysics at Göttingen University since 1996, and teaches courses on plate tectonics and exploration geophysics, electromagnetic depth sounding, mixing laws, and introduction to geo- and astrophysics. Professor Bahr's research interests in addition to MT include percolation theory and conduction mechanisms. He has been an editor of *Geophysical Journal International* since 1997.

Practical Magnetotellurics

Fiona Simpson
and
Karsten Bahr
Georg – August – Universität,
Göttingen

CAMBRIDGE
UNIVERSITY PRESS

University Printing House, Cambridge CB2 8BS, United Kingdom

One Liberty Plaza, 20th Floor, New York, NY 10006, USA

477 Williamstown Road, Port Melbourne, VIC 3207, Australia

314-321, 3rd Floor, Plot 3, Splendor Forum, Jasola District Centre, New Delhi - 110025, India

79 Anson Road, #06-04/06, Singapore 079906

Cambridge University Press is part of the University of Cambridge.

It furthers the University's mission by disseminating knowledge in the pursuit of
education, learning and research at the highest international levels of excellence.

www.cambridge.org
Information on this title: www.cambridge.org/9781108462556

First published 2005
First paperback edition 2018

A catalogue record for this publication is available from the British Library

Library of Congress Cataloging in Publication data
Simpson, Fiona, 1969–
   Practical magnetotellurics / Fiona Simpson and Karsten Bahr.
      p.   cm.
   Includes bibliographical references and index.
   ISBN 0 521 81727 7 (hardback)
   1. Magnetotelluric prospecting.   I. Bahr, Karsten, 1956–   II. Title.

TN269.S5332 2005
622´.153–dc22

ISBN 978-0-521-81727-1 Hardback
ISBN 978-1-108-46255-6 Paperback

There are two possible outcomes: If the result confirms the hypothesis, then you've made a measurement. If the result is contrary to the hypothesis, then you've made a discovery.

Enrico Fermi

Contents

Preface

This book was written for students and researchers in geophysics, geology, and other Earth sciences, who wish to apply or understand the magnetotelluric (MT) method. It is intended to be an introduction to the subject, rather than an exhaustive treatise. At the same time, we do not shirk raising controversial issues, or questions for which there are no easy answers, as we do not wish to give the impression that all of the interesting problems have been solved. MT is very much a dynamic, evolving science.

We acknowledge a bias towards long-period MT studies of the deep crust and mantle. Just as one cannot drink the water from the bottom of a glass until one has drunk the water from the top (unless one has a drinking straw), electromagnetic waves cannot penetrate the deep crust or mantle without being influenced by overlying crustal structures. Hence, longer-period electromagnetic waves, which penetrate deeper into the Earth than shorter-period waves, will necessarily image a higher level of complexity than shorter-period waves. The student who has understood long-period MT sounding should, therefore, have no problem applying their knowledge to audiomagnetotellurics (AMT) and shallow crustal studies.

We have organised the chapters according to the sequence of steps most likely to be encountered by a student embarking on an MT project: from theory to field campaign, to data processing and modelling, through to tectonic and geodynamic interpretation. Some mathematical tools and derivatives are included in the Appendices.

All subjects of a scientific or technical nature have a tendency to spawn jargon, and MT is no exception. Words or phrases that may be deemed jargon are highlighted in italics, and are explained in a Glossary.

No man (or woman) is an island. We extend special thanks to Rainer Hennings who helped with illustrations.

Symbols

Symbols for which no units are given refer to dimensionless parameters. Note that this list does not include symbols used in the
Appendices. We have endeavoured to use the symbols most commonly assigned to common parameters. Occasionally, this results in
a symbol having more than one meaning. Where ambiguity occurs,
the chapter in which a symbol has a different meaning than the
meaning that occurs more frequently in the book is noted in
parentheses.

$\underline{\underline{A}}$	local anisotropy operator
A, B, C, E	impedance commutators $[\mathrm{V^2\,A^{-2}}]$
$\mathbf{B}$	magnetic field, magnetic induction
	$\left[\mathrm{Tesla\ (T) = V\,s\,m^{-2}}\right]$
B_x, B_y, B_z	components of $\mathbf{B}$ in Cartesian co-ordinates [T]
B_r, B_ϑ, B_λ	components of $\mathbf{B}$ in spherical co-ordinates [T]
C	Schmucker–Weidelt transfer function [km]
$\underline{\underline{C}}$	local scatterer distortion tensor (Chapter 5)
$c_{11}, c_{12}, c_{21}, c_{22}$	elements of $\underline{\underline{C}}$
d	distance [km]
$\mathbf{D}$	electric displacement $\left[\mathrm{C\,m^{-2} = A\,s\,m^{-2}}\right]$
D_1, D_2, S_1, S_2	modified impedances $\left[\mathrm{V\,A^{-1}}\right]$
$\mathrm{d}t$	sampling interval [s]
$\mathbf{E}$	electric field $\left[\mathrm{Vm^{-1}}\right]$
E_x, E_y	components of $\mathbf{E}$ in Cartesian components $\left[\mathrm{Vm^{-1}}\right]$
$\tilde{E}$	electric east–west component in the frequency domain $\left[\mathrm{Vm^{-1}}\right]$
f	frequency [Hz]
g	scalar galvanic factor (Chapter 5)
g^{i}, g^{e}	spherical harmonic expansion coefficients (internal and external parts) [T]
g_{ik}	spectral window weight (Chapter 4)
h	surface-layer thickness [km]

H	magnetic intensity $\left[\mathrm{A\,m^{-1}}\right]$
I	current $[\mathrm{A}]$
j	current density $\left[\mathrm{A\,m^{-2}}\right]$
$\mathbf{k}$	wavenumber $\left[\mathrm{m^{-1}}\right]$
K	capacitance $\left[\mathrm{F = A\,s\,V^{-1}}\right]$
l	layer thickness $[\mathrm{km}]$
M_R	model roughness
$\tilde{N}$	electric north–south component in the frequency domain $\left[\mathrm{V\,m^{-1}}\right]$
p	skin depth, penetration depth $[\mathrm{km}]$
$P_n^m(\vartheta)$	associated Legendre polynomials
q	inverse homogenous half-space model transfer function $\left[\mathrm{km^{-1}}\right]$
$\underline{\underline{Q}}$	magnetic field distortion tensor $\left[\mathrm{A\,V^{-1}}\right]$
R	resistance $\left[\Omega = \mathrm{V\,A^{-1}}\right]$
r, ϑ, λ	spherical co-ordinates
S	sensitivity $\left[\mathrm{mVnT^{-1}}\right]$
S_1, S_2, D_1, D_2	modified impedances $\left[\mathrm{V\,A^{-1}}\right]$
$\underline{\underline{S}}$	local shear operator (Chapter 5)
t	time $[\mathrm{s}]$
T	period $[\mathrm{s}]$
$\underline{\underline{T}}$	local twist operator (Chapter 5)
T_x, T_y	induction arrow components
U	scalar potential of **B** (Chapter 1) $\left[\mathrm{Vs\,m^{-1}}\right]$
U	general field vector (Chapter 6)
U	voltage (Chapters 3 and 8) $[\mathrm{V}]$
v	velocity $\left[\mathrm{m\,s^{-1}}\right]$
$w(f)$	frequency-dependent convolution function describing sensor sensitivity (Chapter 4)
w_i	weight of the ith frequency in a robust processing scheme
$\underline{\underline{W}}$	perturbation matrix (Chapter 10)
x, y, z	Cartesian co-ordinates (z positive downwards)
$\tilde{X}, \tilde{Y}, \tilde{Z}$	magnetic north, east and vertical components in the frequency domain $\left[\mathrm{A\,m^{-1}}\right]$
z	depth $[\mathrm{km}]$
Z	impedance $\left[\mathrm{V\,A^{-1}}\right]$
Z_n	impedance of the nth layer of a layered-Earth model $\left[\mathrm{V\,A^{-1}}\right]$
$\underline{\underline{Z}}$	impedance tensor $\left[\mathrm{V\,A^{-1}}\right]$
$Z_{xx}, Z_{xy}, Z_{yx}, Z_{yy}$	elements of $\underline{\underline{Z}}$ $\left[\mathrm{V\,A^{-1}}\right]$
$Z^\mathrm{V}, Z^\mathrm{D}$	upwards, downwards biased estimate of the impedance $\left[\mathrm{V\,A^{-1}}\right]$

α	rotation angle [°]
α	smoothing constant in the non-dimensional weight function (Equation 7.7) used to parameterise model roughness in 2-D *RRI inversion* (Chapter 7)
β	probability
$\underset{=\alpha}{\beta}$	rotation matrix
χ^2	Groom–Bailey misfit measure of the 2-D model with local scatterer
γ	electrical connectivity
δ	phase difference between the elements in one column of $\underline{Z}$ [°]
δE	electric noise (Chapter 4) $\left[\mathrm{V\,m^{-1}}\right]$
$\delta\phi$	phase difference between the principal polarisations of $\underline{Z}$ [°] (Chapter 5)
$\varepsilon, \varepsilon_0$	electrical permittivity, electrical permittivity of free space $\left[\mathrm{A\,s\,V^{-1}\,m^{-1} = F\,m^{-1}}\right]$
ε^2	model misfit measure
ε^2	residuum (Chapters 3, 4)
η_{f}	electric charge density (Chapter 2) $\left[\mathrm{C\,m^{-3} = A\,s\,m^{-3}}\right]$
η	misfit measure for the 2-D model with local scatterer (phase-sensitive skew)
η	constant in the non-dimensional weight function (Equation 7.7) used to parameterise model roughness in 2-D *RRI inversion* (Chapter 7)
η	porosity (Chapter 8)
ϑ, λ, r	spherical co-ordinates
κ	2-D model misfit measure (Swift skew)
λ	wavelength [m]
μ, μ_0	magnetic permeability, magnetic permeability of free space $\left[\mathrm{V\,s\,A^{-1}\,m^{-1} = H\,m^{-1}}\right]$
μ	misfit measure of the layered-Earth model with local scatterer (Chapter 5)
ν	number of degrees of freedom
ρ	resistivity $\left[\Omega\,\mathrm{m} = \mathrm{V\,m\,A^{-1}}\right]$
ρ_{a}	apparent resistivity $\left[\Omega\,\mathrm{m} = \mathrm{V\,m\,A^{-1}}\right]$
σ	conductivity $\left[\mathrm{S\,m^{-1} = A\,V^{-1}\,m^{-1}}\right]$
σ	variance (Chapters 4 and 5)
Σ	layered-Earth misfit measure
τ	conductance $\left[\mathrm{S = A\,V^{-1}}\right]$

τ	relaxation time (Chapter 8, Section 8.1)
ϕ	magnetotelluric phase [°]
ϕ	volume fraction of the conductive phase in two-phase media (Chapter 8)
ω	angular frequency $[\text{s}^{-1}]$
ψ	coherence

Chapter 1

Introduction

Magnetotellurics (MT) is a passive exploration technique that utilises a broad spectrum of naturally occurring geomagnetic variations as a power source for electromagnetic induction in the Earth. As such, MT is distinct from active geoelectric techniques, in which a current source is injected into the ground as a power source for conduction. In fact, MT and geoelectrics have little in common other than the physical parameter (electrical conductivity) imaged. MT is more closely related to geomagnetic depth sounding (GDS), which was developed in the late nineteenth century after the existence of magnetovariational fields arising from induction was demonstrated by Schuster (1889) and Lamb (see Schuster, 1889, pp. 513–518). They applied a mathematical technique, invented by Gauss (1839) for separating magnetovariational fields originating internal to the Earth from those of external origin, to geomagnetic observatory data and detected a significant internal component. In the 1950s, Tikhonov (1950, reprinted 1986) and Cagniard (1953) realised that if electric and magnetic field variations are measured simultaneously then complex ratios (impedances) can be derived that describe the penetration of electromagnetic fields into the Earth. The penetration depths of electromagnetic fields within the Earth depend on the electromagnetic sounding period, and on the Earth's conductivity structure. This is the basis of the MT technique.

1.1 Magnetotellurics as a passive electromagnetic exploration method and its relation to active electromagnetic and geoelectric methods

The magnetotelluric (MT) technique is a passive electromagnetic (EM) technique that involves measuring fluctuations in the natural electric, **E**, and magnetic, **B**, fields in orthogonal directions at the surface of the Earth as a means of determining the conductivity structure of the Earth at depths ranging from a few tens of metres to several hundreds of kilometres. The fundamental theory of exploration MT was first propounded by Tikhonov (1950, reprinted 1986) and, in more detail, by Cagniard (1953). Central to the theses of both authors was the realisation that electromagnetic responses from any depth could be obtained simply by extending the magnetotelluric sounding period. This principle is embodied in the *electromagnetic skin depth* relation, which describes the exponential decay of electromagnetic fields as they diffuse into a medium:

$$p(T) = (T/\pi\mu\bar{\sigma})^{1/2}, \tag{1.1}$$

where $p(T)$ is the electromagnetic skin depth in metres at a given period, T, $\bar{\sigma}$ is the average conductivity of the medium penetrated, and μ is magnetic permeability. At a depth, $p(T)$, electromagnetic fields are attenuated to e^{-1} of their amplitudes at the surface of the Earth. This exponential decay of electromagnetic fields with increasing depth renders them insensitive to conductivity structures lying deeper than $p(T)$. Hence, in MT studies, one electromagnetic skin depth is generally equated with the *penetration depth* of electromagnetic fields into the Earth. In studies of the Earth, μ is usually assigned the free-space value ($\mu_0 = 4\pi \times 10^{-7}\,\mathrm{H\,m^{-1}}$), and Equation (1.1) can be approximated as

$$p(T) \approx 500\sqrt{T\rho_a}, \tag{1.2}$$

where ρ_a is *apparent resistivity*, or the average resistivity of an equivalent uniform *half-space*.

From Equations (1.1) and (1.2), we can deduce that for a given sounding period, the depth achieved by passive EM sounding will be dictated by the average conductivity of the overlying sample of earth that is penetrated. Electromagnetic fields that are naturally induced in the Earth and are exploitable for MT studies have wave periods ranging from $\sim 10^{-3}$ to $\sim 10^5$ s. Therefore, if we assume an average resistivity of the Earth's crust and upper mantle of $100\,\Omega\,\mathrm{m}$ (we hasten to add that the conductivity structure of the Earth is much more interesting!), we can see how penetration depths in the range of ~ 160 m to >500 km might be possible. The broad span of depths that can be imaged using the MT technique is one advantage of the method compared with active EM methods for which the maximum depth that can be probed is always limited by the size of the available source, and realisable source–receiver configurations.

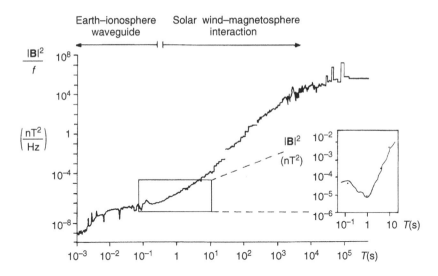

Figure 1.1 Power spectrum illustrating '1/f characteristics' of natural magnetic variations (modified from Junge, 1994). Short-period signals are generated by interactions in the Earth–ionosphere waveguide, whereas long-period signals are generated by solar wind–magnetosphere interactions. The spectral lines at periods of the order 10^5 s are harmonics of the solar quiet (Sq) daily variation. Inset illustrates the reduced signal power ($|\mathbf{B}|^2$) in the dead-band.

Whilst magnetohydrodynamic processes within the Earth's outer core generate the greater part of the Earth's magnetic field, it is the superimposed, more transient, lower-amplitude fluctuations of external origin, that MT sounding seeks to exploit. The power spectrum (Figure 1.1) of these fluctuations plummets in the 0.5–5 Hz frequency range, minimising at a frequency of ~1 Hz. This so-called *dead-band* of low-amplitude signals is attributable to the inductive source mechanisms, one effective above ~1 Hz, the other below ~1 Hz, and is frequently manifest in MT sounding curves by a reduction in data quality.

Electromagnetic fields with frequencies higher than 1 Hz (i.e., periods shorter than 1 s) have their origins in meteorological activity such as lightning discharges. The signals discharged by lightning are known as 'sferics' and encompass a broad range of electromagnetic frequencies. Local lightning discharges may saturate amplifiers, and it is not these, but rather the sferics from the highly disturbed equatorial regions, which propagate around the world within the waveguide bounded by the ionosphere and Earth's surface that are of most significance. Sferics propagate within the waveguide as transverse electric (TE), transverse magnetic (TM) or transverse electric and magnetic (TEM) waves, and are enhanced or attenuated depending on frequency. A description of these waveguide modes and discussion of the constructive and destructive interference that leads to certain frequencies being enhanced, whilst others are attenuated can be found in Dobbs (1985, Chapter 8). During the day, the waveguide is ~60 km wide, increasing to ~90 km at night-time. Sferics peak in the early afternoon. However, statistically a part of the

Figure 1.2 (a) Distortion of advancing front of solar plasma by Earth's magnetic field (after Chapman and Ferraro, 1931). The solar wind blows continually, but with varying intensity. Sudden increases in solar-wind pressure push the *magnetopause* closer to the Earth giving rise to magnetic storms. (b) Magnetic field lines showing the form of the Earth's magnetosphere, which extends further from the Earth than originally envisaged by Chapman and Ferraro (1931). The magnetosphere typically extends to approximately 10 Earth radii (i.e., ~64 000 km) on the dayward side of the Earth, whilst a long magnetic tail (the magnetotail) extends more than 300 000 km away from the nightward side of the Earth.

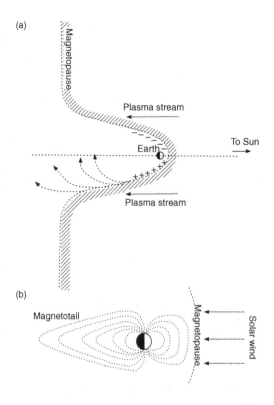

world may witness thunderstorm activity at any given universal time (UT).

Interactions between the solar wind (Parker, 1958) and the Earth's *magnetosphere* and ionosphere generate electromagnetic fluctuations with frequencies lower than 1 Hz (i.e., periods longer than 1 s). Briefly, the solar wind is a continual stream of plasma, radiating mainly protons and electrons from the Sun. On encountering the terrestrial magnetic field (at the *magnetopause*), these protons and electrons are deflected in opposite directions, thereby establishing an electric field (Figure 1.2). Variations in density, velocity and magnetic field intensity of the solar wind produce rapidly varying distortions of the Earth's *magnetosphere*. For example, increases in solar-wind pressure cause rapid compression of the magnetosphere, and therefore compaction of magnetic field lines, affecting an increase in the horizontal geomagnetic field. Oscillations of the magnetosphere generate small, almost sinusoidal variations of the geomagnetic field, called geomagnetic pulsations. Inductive and magnetohydrodynamic interactions between the magnetosphere and ionosphere complexly modify these fluctuating fields before they reach the Earth's surface. A more detailed discussion of these processes can be found in Campbell (1997).

The largest geomagnetic field variations (up to the order of a few hundred nT) occur during *magnetic storms*, which occur owing to sporadic increases in the rate at which plasma is ejected from the Sun. Magnetic storms last for several days, and in polar regions can lead to magnificent displays of light known as *aurora borealis* and *aurora australis*, or *northern* and *southern lights*, respectively.

Active EM induction techniques, in which the EM source field is man-made (e.g., the output from a generator), probe a more limited range of penetration depths than passive EM induction techniques: although, in theory, an artificial EM field with a period of several days can be generated, such a field will not be as powerful as a magnetic storm. Hence, MT sounding is better suited to probing depths of several hundred kilometres than are active induction techniques. However, active EM induction techniques can be used to improve the data quality in the period range of the dead-band.

An active exploration technique that is commonly used for probing the Earth's electrical conductivity structure is geoelectrics. Here, the physical process involved is conduction – which is not time-dependent – rather than induction. The spatial distribution of the geoelectrical voltage is described by the Laplace equation (rather than by the time-dependent *diffusion equation* that governs induction, which we explore in detail in Chapter 2): i.e., geoelectrics is a potential method. Being a potential method, geoelectrics suffers from the limitations which all potential methods have: in particular, a very limited depth resolution. In contrast, MT is not a potential method, and if EM fields spanning a wide period range are evaluated, the depth resolution of the MT method can be very high.

1.2 Problems for which EM studies are useful: a first overview of conduction mechanisms

Ascertaining subterranean electrical resistivity structure is rendered trivial unless electrical resistivity can be linked to other physical parameters or processes. Temperature, pressure, physical and chemical state, *porosity* and *permeability*, as well as the frequency at which measurements are made can all play a crucial role in determining the electrical resistivity exhibited by rocks and minerals.

The transmission of electrical currents by free charge carriers is referred to as *conduction*. Conduction occurs in rocks and minerals via three principal electrical charge propagation mechanisms: electronic, semi-conduction and electrolytic. Electronic conduction occurs in metallic ore minerals (e.g., magnetite, haematite), which contain free electrons that can transport charge. Electronic conduction is most significant in the Earth's core. Conduction in graphite is of particular relevance in EM studies. A single-crystal of graphite is an electronic conductor, due to the availability of free electrons. The conductivity of the amorphous graphite that occurs in the Earth's crust is lower than the conductivity of metallic ores, but higher than the conductivity of natural semi-conductors or electrolytes. Semi-conduction occurs in poor conductors containing few free charge carriers. Only a small proportion of the electrons present in a semi-conductor contribute to conduction, and a marked increase in conduction can be affected by the inclusion of minor amounts of weakly-bonded, electron-donating impurities. Semi-conduction is expected to dominate in mantle minerals such as olivine. Electrolytic conduction occurs in a solution containing free ions. Saline water is an electrolyte of particular relevance to crustal EM studies. Its free ions are derived from the dissolution of the constituent ions (e.g., Na^+, Cl^-) of the solid salt on entering solution. Only 'free' (i.e., not chemically bound) water has a significant effect on observed conductivities. In active tectonic regions, any partial melt generated by enhanced temperatures, adiabatic decompression or asthenospheric upwelling will also act as an electrolyte (e.g., Shankland and Waff, 1977).

EM sounding methods are highly sensitive to variations in abundance and distribution of minor mineral constituents and fluids within the rock matrix. A small fraction of conductive mineral (or fluid) will increase the *bulk conductivity* of a rock formation considerably if distributed so that it forms an interconnected network, whereas a larger fraction of the same conductive phase could have negligible effect if present as isolated crystals (or in pores). The interdependence of quantity and distribution of a conductive phase in determining the conductivity of a multi-phase medium is described by mixing laws (see Section 8.3). Conductive minerals that occur within the Earth in the required abundance and geometries to significantly affect the electrical resistivities include graphite, sulphides, magnetite, haematite, pyrite and pyrrhotite.

As we shall see later (Chapter 8), bulk electrical conductivity is relatively insensitive to the electrical conductivity of the host rock. Therefore, a direct link between electrical conductivity and lithology is generally not possible. However, porosity and permeability vary significantly according to rock type and formation, and variations in porosity and permeability influence electrical conductivity substantially. Sedimentary rocks typically contain significant interstitial fluid content, and are generally more porous than igneous or metamorphic rocks, unless the latter are extensively fractured, fissured or jointed. Unconsolidated sedimentary rocks can support porosities of up to 40%, but older, more deeply buried (and therefore more densely compacted) sediments might have typical porosities of less than 5%. In limestones and volcanic rocks, spheroidal pores (known as vugs) are common, but fluids contained in vugs will have less effect on bulk rock conductivity than fluids distributed in thin, interconnected microcracks or as grain-boundary films. Whereas the conductivity of a saturated rock decreases significantly with pressure owing to closure of fluid-bearing cracks, that of a partially saturated rock may be enhanced for small pressure increments, as a result of new fluid pathways being generated by microcracking. In addition to conduction via pores and cracks, adsorption of electrically attracted ions at grain interfaces results in surface conduction. Surface conduction is of particular significance in clays, which ionise readily.

The resistivity of a two-phase medium is governed primarily by the connectivity and resistivity of the conductive phase. Fluid resistivities are sensitive to salinity, pressure and temperature. With increasing temperature, the viscosity of a solution is decreased, promoting heightened ionic mobility and therefore tending to enhance conductivity. However, there are also conflicting tendencies owing to decreasing density and decreasing dielectric constant with increasing temperature

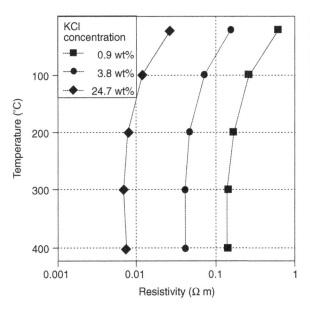

Figure 1.3 Electrical resistivities of varying concentrations of aqueous KCl solution as a function of temperature. (Redrawn from Nesbitt, 1993.)

that act to reduce conductivity. Figure 1.3 shows resistivity as a function of temperature for high-salinity KCl brines. It is known from boreholes that highly saline fluids exist to at least mid-crustal depths. For example, fluids pumped from depths down to 8.9 km from a deep borehole (KTB) in Bavaria, S. Germany have NaCl salinities of 1.7 M (Huenges *et al.*, 1997), which translates to electrical resistivities of $\sim$0.1 Ω m at 20 °C, or $\sim$0.025 Ω m at 260 °C – the temperature at the 8.9 km depth from which they were pumped. Seawater has electrical resistivities of the order 0.2–0.25 Ω m.

A major contender to fluids as the cause of enhanced crustal conductivities is carbon – either mantle-derived as a product of CO_2 degassing (Newton *et al.*, 1980) or biogenic (e.g., Rumble and Hoering, 1986; Jödicke, 1992) – in the form of graphite and/or black shales. It has been suggested that, under strongly reducing conditions, graphite might form as a grain-boundary precipitate from CO_2-rich or hydrocarbon-bearing metamorphic fluids (Frost, 1979; Glassley, 1982; Rumble and Hoering, 1986 and references therein). CO_2 inclusions in metasedimentary granulites (which may be representative of lower-crustal rock type) from Scandanavia, Tanzania and India are consistent with mantle-derived, CO_2-rich fluids and carbonate-rich melts emplaced by deep-seated intrusives (Touret, 1986). However, C_{13} isotope studies support a biogenic source of graphite in outcrops that have been sampled (e.g., Rumble and Hoering, 1986, Large *et al.*, 1994). In contrast to artificial manufacture of graphite via carbon vapour diffusion for which formation temperatures of between 600 °C (with nickel as a catalyst) and greater than 2000 °C are required, graphite has been identified in low-grade metamorphic rocks with inferred formation temperatures of less than 450 °C (e.g., Léger *et al.*, 1996). The low formation temperatures may be explained by the action of shear stresses, which promote graphitisation in the 400–500 °C temperature range (Ross and Bustin, 1990).

Graphitic horizons have been interpreted as markers of past tectonic events. For example, Stanley (1989) documented extensive outcrops of graphitic shales apparently associated with active and fossilised subduction zones, and exposures of black shales have also been mapped along the Grenville Front (Mareschal *et al.*, 1991). Graphitic films have been observed in anorthosite rocks derived from the mid crust in Wyoming (Frost *et al.*, 1989), whilst core samples from the KTB borehole and proximate surface exposures contain graphite both as disseminated flakes and concentrated along shear zones (Haak *et al.*, 1991; ELEKTB, 1997). Graphite has also been detected in mafic xenoliths of recent crustal origin

(Padovani and Carter, 1977). During uplift, loss of connectivity of the graphite may occur (Katsube and Mareschal, 1993; Mathez *et al.*, 1995), but Duba *et al.* (1994) and Shankland *et al.* (1997) have demonstrated that reconnection of graphite occurs when rock samples are pressurised. This pressure-induced reconnection of graphite suggests a rock mechanical model by which rocks, which are resistive at the Earth's surface may be conductive at depth. An alternative explanation involves loss of fluids (Shankland and Ander, 1983). Arguments for and against saline fluids and/or graphite as prime candidates for generating enhanced conductivities in the deep continental crust have been reviewed by Yardley and Valley (1997, 2000), Simpson (1999) and Wannamaker (2000).

Long-period MT studies (e.g., Lizzaralde *et al.*, 1995, Heinson, 1999, Simpson, 2001b) indicate that average mantle conductivities are higher than the conductivity of dry olivine (Constable *et al.*, 1992), which constitutes 60–70% of the upper mantle. Enhanced mantle conductivities may be generated by graphite, partial melts or diffusion of free ions (e.g., H^+, OH^-).

Partial melting may account for deep-crustal and upper-mantle conductivity anomalies that are imaged coincident with enhanced heat flow and seismic low-velocity zones (e.g., Wei *et al.*, 2001). The conductivity of a partial melt is governed by its composition and its temperature (Figure 1.4), as well as by melt fraction and distribution (see Chapter 8). Sometimes, however, modelled electrical conductances imply implausibly high mantle temperatures or melt fractions for them to be generated solely by partial melting.

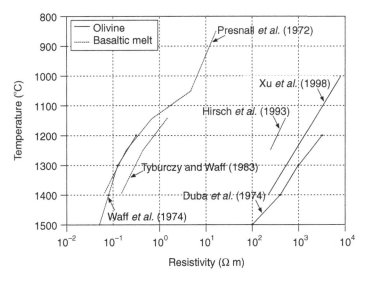

Figure 1.4 Electrical resistivities of dry olivine and basaltic melt as a function of temperature. Basaltic melt is approximately three orders of magnitude less resistive than dry olivine.

Trace amounts (p.p.m.) of water can significantly enhance the electrical conductivity of olivine at a given temperature by supplying mobile charge carriers in the form of hydrogen (H^+) ions (Karato, 1990). Hydrogen diffusivities are anisotropic in olivine (Mackwell and Kohlstedt, 1990), and the implications of this in terms of the potential contributions that MT studies can make to geodynamic models of mantle flow will be discussed in Chapter 9 (Section 9.3).

The maximum penetration depths that can be routinely achieved using MT sounding are of the order 400–600 km. At these depths, phase transitions from olivine to wadsleyite (~410 km) to ringwoodite (~520 km) to perovskite + magnesiowüstite (~660 km) are believed to occur. These phase transitions are expected to contribute to significant conductivity increases with increasing penetration into the *transition zones* (Xu et al., 1998). Conductivity–depth models for the mantle derived from long-period MT data and from laboratory conductivity measurements on representative mineral assemblages are compared in more detail in Chapter 8 (Sections 8.1 and 8.2). Seismological studies also image transition zones at mid-mantle depths of ~410 km, ~520 km and ~660 km.

Overall, the electrical resistivities of rocks and other common Earth materials span 14 orders of magnitude (Figure 1.5). Dry crystalline rocks can have resistivities exceeding 10^6 Ω m, whilst graphite-bearing shear zones might have resistivities of less than 0.01 Ω m. Such a wide variance provides a potential for producing well-constrained models of the Earth's electrical conductivity structure.

1.3 An historical perspective

In 1889, Schuster (1889) applied the *Gauss separation* technique to the *magnetic daily (diurnal) variation*, and the subject of electromagnetic induction in the Earth was born. The Gauss separation is a mathematical formulation used by Gauss (1839) to deduce that the origin of the geomagnetic main field is internal. Suppose we have competition between two magnetic field models – a dipole within the Earth and a large ring current outside the Earth (Figure 1.6). In both cases, the magnetic field vector is the gradient of a potential:

$$\mathbf{B} = -\nabla U \qquad (1.3)$$

and in both cases the components B_r, B_ϑ (in spherical co-ordinates, at the Earth's surface) are the derivatives of the potential U with respect to the co-ordinates r, ϑ, at the location $r = r_E$, where r_E is the radius of the Earth. For the dipole model

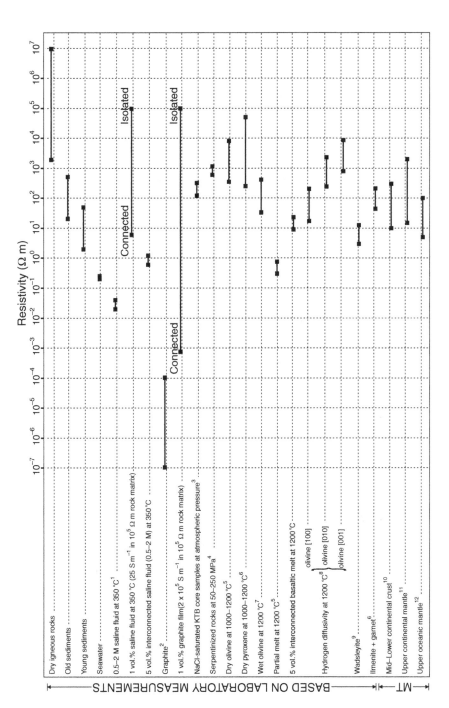

Figure 1.5 Electrical resistivities of rocks and other common Earth materials. The labels 'connected' and 'isolated' refer to the more conductive component (saline fluid/graphite/basaltic melt) of a two-phase medium, and the values are calculated from Hashin–Shtrikman (1962) upper and lower bounds (see Chapter 8), assuming a rock matrix with a resistivity of $10^5 \, \Omega \, \mathrm{m}$.

References: [1]Nesbitt, 1993; [2]Duba and Shankland, 1982; [3]Duba *et al.*, 1994; [4]Stesky and Brace, 1973; [5] see Figure 1.4; [6]Xu and Shankland, 1999 (and references therein); [7]Lizzaralde *et al.*, 1995, and references therein; [8]Karato, 1990, and Kohlstedt and Mackwell, 1998; [9]Xu *et al.* 1998; [10] Haak and Hutton, 1986, and Jones, 1992; [11]Heinson and Lilley, 1993, and Lizzaralde *et al.*, 1995; [12]Schultz *et al.*, 1993, and Simpson, 2002b.

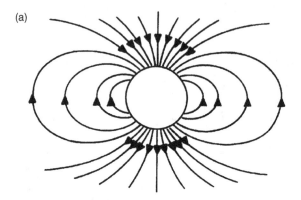

(a)

Figure 1.6 Magnetic field due to (a) geocentric dipole, (b) ring current encircling the Earth.

(b)

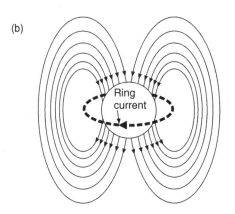

$$U = \frac{g_1^i r_E^3 \cos \vartheta}{r^2}, \qquad B_r = 2g_1^i \cos \vartheta, \qquad B_\vartheta = g_1^i \sin \vartheta, \qquad (1.4)$$

and for the ring-current model

$$U = r g_1^e \cos \vartheta \qquad B_r = -g_1^e \cos \vartheta, \qquad B_\vartheta = g_1^e \sin \vartheta, \qquad (1.5)$$

where g_1^i and g_1^e are the first-degree spherical harmonic expansion coefficients for the internal and external parts, respectively, of the field. We assume that the diameter of the ring current is so large that it creates an homogeneous field at the surface of the Earth. We see from a comparison of Equations (1.4) and (1.5) or directly from Figure 1.6. that the form of the magnetic field is different for these two models, and that the *inclination, I*:

$$I = \arctan(B_r/B_\vartheta) \qquad (1.6)$$

varies along a path from the equator to the pole in different ways for the two models. Adding Equations (1.4) and (1.5) we can generate a model with internal and external parts, or if the components

B_r and B_ϑ or the inclination are available we can distinguish between the two models. In this way, Gauss deduced that the external part of the geomagnetic main field can be neglected.[1] On the other hand, Schuster's (1889) application of the Gauss separation technique to the magnetic daily variation revealed a dominant external part.

What is the magnetic daily variation? In the eighteenth century, it was known that the *declination* exhibits a daily variation with an amplitude of 0.2° (Graham, 1724). Later, Gauss deduced from the measurements made by his 'Göttingen magnetic society', that irregular magnetic variations with periods shorter than 1 day exist, and that these variations are correlated within a Europe-wide network of declination measurements (Gauss, 1838). The origin of these variations was not known, but there was some speculation about 'electricity in the air' (Gauss, 1838). Today, we know that thermal tides generate current vortices in the ionosphere, and the daily magnetic variation is the magnetic field caused by these currents (e.g., Parkinson, 1971; Winch, 1981).[2]

Although Schuster (1889) demonstrated that the magnetic daily variation is predominantly of external origin, a significant internal part was also apparent in the Gauss separation. In an appendix to Schuster's paper, Lamb (Schuster, 1889, pp. 513–518) showed that this internal part is due to induction in the conductive Earth, and Schuster (1889, p. 511) deduced from the phase of the internal part that the average conductivity of the Earth increases with depth. This was confirmed in subsequent studies by Chapman (1919) and Lahiri and Price (1939). In particular, a model of Lahiri and Price (1939) has a conductivity jump at 600 km, anticipating our modern knowledge that the first harmonic of the magnetic daily variation penetrates to depths of the order of the transition zones in the mid mantle.

Thus far, we have not mentioned electric field measurements – but isn't MT a combination of magnetic and electric ('telluric') variational measurements? Yes, but we wanted to emphasise that MT (the theory of which was propounded independently by Tikhonov (1950) and Cagniard (1953)) is not the only passive, large-scale geophysical technique that utilises induction in the

---

[1] Gauss allowed for more complicated models by expanding the potential into spherical harmonics: $g_1{}^i$ and $g_1{}^e$ are the first coefficients of the internal and external part in that expansion.

[2] The form of these current systems is more complicated then the large ring current in Figure 1.6, and spherical harmonics cannot be avoided if we want to describe their form precisely (see Section 10.2).

Earth. Whereas, for the potential-separation technique, a world-wide site coverage or knowledge of the form of the ionospheric current vortices is necessary, MT has the advantage that measurements at a single site can be readily interpreted (as a consequence of the *plane wave* assumption – see Section 2.2). However, MT practitioners must be wary of distortions in electric fields (Chapter 5), which are not present in magnetovariational techniques. The two techniques lead to complementary results (Schmucker, 1973), as we shall discuss in Section 10.2 and Appendix 2.

1.4 MTnet

MTnet is an internet forum for the free exchange of programs, data and ideas between scientists engaged in non-commercial studies of the Earth using EM techniques, principally MT. The address of the MTnet web page is www.mtnet.info

1.5 Books that cover other aspects of EM

There are many good textbooks on applied geophysics, and most of these devote a chapter to EM methods (e.g., Reynolds, 1997). Passive EM methods, utilising natural sources, often get a short mention, but active, controlled source techniques dominate in applied geophysics texts.

As we shall see in Chapter 2, an in-depth knowledge of the physics of the processes in the Earth's ionosphere and magnetosphere that result in natural time-varying electromagnetic fields is not necessary in order to apply MT methods. Readers who nevertheless want to know more about these processes are referred to the textbook by Campbell (1997).

In Chapters 6 and 7, we treat the subject of numerical modelling. The intention in these chapters is to provide practical guidance in using existing forward and inverse modelling algorithms, rather than a comprehensive coverage of the numerical methods employed. Those researchers who wish to write their own codes are referred to the excellent book *Mathematical Methods in Electromagnetic Induction* by Weaver (1994).

Chapter 2
Basic theoretical concepts

Starting from Maxwell's equations, we derive the diffusion equation for the electric field. Diffusion governs not only electromagnetic induction but also the spreading of thermal fields. Therefore, there is an analogue in your kitchen: pre-heat the oven to 250 °C, and put a roast (2 kg) into it. Remove after 15 minutes and cut in two halves. The outermost 2 cm are cooked but the inner bit is quite raw. This tells us that 2 cm is the penetration depth of a thermal field with period 15 min in beef. A shorter period would yield a smaller penetration depth and a longer period would penetrate deeper. In addition, the thermal field arrives with a delay inside the beef. This delay is governed by the thermal conductivity of the beef. Electromagnetic induction in the earth is governed by the skin effect and behaves in a similar way to the beef analogue: there is a period-dependent penetration depth, and we observe the delayed penetration (phase lag) and decay of an electromagnetic field into the conductive subsoil. Due to the phase lag of the penetrating fields we have a complex penetration depth, which we call a 'transfer function'. Electromagnetic sounding is a volume sounding. Therefore, for the simplest case of an homogeneous Earth, MT transfer functions contain information about the electrical conductivity in a hemisphere, with the magnetotelluric site located at the centre of the bounding horizon. Some mathematical gymnastics allow us to write Maxwell's equations either in spherical co-ordinates (if the conductivity of the entire Earth is the subject), or in Cartesian co-ordinates if a regional survey is performed. Because of Biot–Savart's Law, we expect that electric and magnetic fields are perpendicular – north–south electric fields should be associated with east–west magnetic fie lds, and vice versa. The impedance tensor is introduced as a first example of hypothesis testing, and it will turn out that the expectation of orthogonal electric and magnetic fields is violated very often.

2.1 Assumptions of the MT method

For the purposes of considering electromagnetic induction in the Earth, a number of simplifying assumptions are considered applicable (e.g., Cagniard, 1953; Keller and Frischknecht, 1966):

(i) Maxwell's general electromagnetic equations are obeyed.

(ii) The Earth does not generate electromagnetic energy, but only dissipates or absorbs it.

(iii) All fields may be treated as conservative and analytic away from their sources.

(iv) The natural electromagnetic source fields utilised, being generated by large-scale ionospheric current systems that are relatively far away from the Earth's surface, may be treated as uniform, plane-polarised electromagnetic waves impinging on the Earth at near-vertical incidence. This assumption may be violated in polar and equatorial regions.

(v) No accumulation of free charges is expected to be sustained within a layered Earth. In a multi-dimensional Earth, charges can accumulate along discontinuities. This generates a non-inductive phenomenon known as *static shift*.

(vi) Charge is conserved, and the Earth behaves as an ohmic conductor, obeying the equation:

$$\mathbf{j} = \sigma \mathbf{E}, \tag{2.1}$$

where, $\mathbf{j}$ is total electric current density (in $A\,m^{-2}$), σ is the conductivity of the sounding medium (in $S\,m^{-1}$), and E is the electric field (in $V\,m^{-1}$).

(vii) The electric displacement field is quasi-static for MT sounding periods. Therefore, time-varying displacement currents (arising from polarisation effects) are negligible compared with time-varying conduction currents, which promotes the treatment of electromagnetic induction in the Earth purely as a diffusion process (see Section 2.3).

(viii) Any variations in the electrical permittivities and magnetic permeabilities of rocks are assumed negligible compared with variations in bulk rock conductivities.

2.2 Time invariance as a consequence of the plane wave assumption

For the purposes of the MT technique, we can assume that at mid-latitudes (far away from the complicated current systems generated by the equatorial and auroral electrojets), large-scale, uniform, horizontal sheets of current far away in the ionosphere give rise to *plane waves* normally incident on the surface of the conductive

Earth. A plane wave is one that propagates normal to a plane in which the fields are constant. (For example, unobstructed ocean waves are approximately planar). For plane electromagnetic waves, electric (**E**) and magnetic (**B**) fields with amplitudes at the origin of $\mathbf{E_0}$ and $\mathbf{B_0}$, angular frequency ω (period, $T = 2\pi/\omega$), and wavelength, $\lambda = 2\pi/|\mathbf{k}|$, (where **k** is wavenumber) take the mathematical forms:

$$\mathbf{E} = \mathbf{E_0}e^{i\omega t - kz} \tag{2.2a}$$

$$\mathbf{B} = \mathbf{B_0}e^{i\omega t - kz}. \tag{2.2b}$$

The plane wave assumption is fundamental to the MT technique, because it implies time invariance of the exciting source. As a consequence of time invariance, the *impedance tensor* calculated from the orthogonal electric and magnetic fields (Section 2.9) at any given site should be self-similar regardless of when the fields are recorded (provided, of course, that the electrical conductivity structure of the Earth does not change, and that signal-to-noise ratios are adequate).

Wait (1954) and Price (1962) showed that limitations of the normally incident plane wave assumption arise when the lateral extension of the source field is not significantly larger than the *penetration depth* (see Section 2.4) of electromagnetic fields into the Earth, requiring second-order corrections to be made to Cagniard's (1953) formulation. However, Cagniard's assumptions concerning the source field are generally valid at mid-latitudes for periods less than 10^4 s (Madden and Nelson, 1964, reprinted 1986). Departures from the plane wave assumption occur in polar and equatorial regions owing to instabilities in the forms of the source fields that arise from complexities in the auroral and equatorial electrojet current systems. Depending on the degree of disturbance, departures from the uniform source can sometimes be circumvented by selection of a subset of undisturbed data. For example, night-time data is generally less disturbed than daytime data. More detailed analysis techniques for dealing with source-field disturbances are discussed in Section 10.3.

For periods of less than a day, the Earth's curvature has no significant effect on the plane wave assumption (Srivastava, 1965). Therefore, for the purposes of MT, it is sufficient to consider a flat-Earth model with electromagnetic fields described in Cartesian co-ordinates.

2.3 Why EM speaks volumes: MT as a vector field method

The behaviour of electromagnetic fields at any frequency is concisely described by Maxwell's equations, which for a polarisable, magnetisable medium may be expressed as:

$$\nabla \times \mathbf{E} = -\frac{\partial \mathbf{B}}{\partial t} \tag{2.3a}$$

$$\nabla \times \mathbf{H} = \mathbf{j}_f + \frac{\partial \mathbf{D}}{\partial t} \tag{2.3b}$$

$$\nabla \cdot \mathbf{B} = 0 \tag{2.3c}$$

$$\nabla \cdot \mathbf{D} = \eta_f, \tag{2.3d}$$

where $\mathbf{E}$ is the electric field (in $V\,m^{-1}$), $\mathbf{B}$ is magnetic induction (in T), $\mathbf{H}$ is the magnetic intensity[3] (in $A\,m^{-1}$), $\mathbf{D}$ is electric displacement (in $C\,m^{-2}$), j_f is the electric current density owing to free charges (in $A\,m^{-2}$) and η_f is the electric charge density owing to free charges (in $C\,m^{-3}$). Curl ($\nabla\times$) and div ($\nabla\cdot$) are *vector calculus* expressions. The most important theorems from vector calculus are listed in Appendix 1.

Equation (2.3a) is Faraday's Law and states that time variations in the magnetic field induce corresponding fluctuations in the electric field flowing in a closed loop with its axis oriented in the direction of the inducing field (Figure 2.1). Equation (2.3b) is Ampère's Law, which states that any closed loop of electrical current will have an associated magnetic field of magnitude proportional to the total current flow. Assuming that time-varying

Figure 2.1 A time-varying external magnetic field, $\mathbf{B}_e$, in accord with Faraday's Law, induces an electric field $\mathbf{E}$, which then induces a secondary, internal magnetic field $\mathbf{B}_i$, in accord with Ampère's Law.

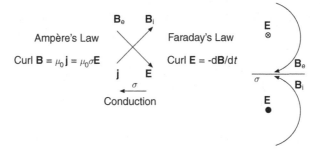

---

[3] Magnetic induction is the force exerted on a charge moving in a magnetic field, whereas magnetic intensity is the magnetising force exerted on a magnetic pole placed in a magnetic field independent of whether or not the pole is moving.

displacement currents are negligible (Section 2.1, assumption (vii))
Ampère's Law reduces to:

$$\nabla \times \mathbf{H} = \mathbf{j}_{\mathrm{f}}. \tag{2.4}$$

Equation (2.3c) states that no free magnetic charges (or mono-
poles) exist. For a linear, isotropic medium, two further relationships have been shown to hold:

$$\mathbf{B} = \mu\mathbf{H} \tag{2.5a}$$

$$\mathbf{D} = \varepsilon\mathbf{E}. \tag{2.5b}$$

For MT studies, variations in electrical permittivities, ε, and
magnetic permeabilities, μ, of rocks are negligible compared with
variations in bulk rock conductivities (Section 2.1, assumption
(viii)), and free-space values ($\varepsilon_0 = 8.85 \times 10^{-12}\,\mathrm{F\,m^{-1}}$ and
$\mu_0 = 1.2566 \times 10^{-6}\,\mathrm{H\,m^{-1}}$) are assumed.

Applying Equations (2.5a) and (2.5b) and Ohm's Law
(Equation (2.1)), Maxwell's equations can be rewritten in the form:

$$\nabla \times \mathbf{E} = -\frac{\partial \mathbf{B}}{\partial t} \tag{2.6a}$$

$$\nabla \times \mathbf{B} = \mu_0 \sigma \mathbf{E} \tag{2.6b}$$

$$\nabla \cdot \mathbf{B} = 0 \tag{2.6c}$$

$$\nabla \cdot \mathbf{E} = \eta_{\mathrm{f}}/\varepsilon. \tag{2.6d}$$

Assuming that no current sources exist within the Earth,

$$\nabla \cdot \mathbf{j} = \nabla \cdot (\sigma\mathbf{E}) = 0. \tag{2.7}$$

For the case of an homogenous *half-space* (i.e., $\nabla\sigma = 0$):

$$\nabla \cdot (\sigma\mathbf{E}) = \sigma\nabla \cdot \mathbf{E} + \mathbf{E}\nabla\sigma = \sigma\nabla \cdot \mathbf{E} \tag{2.8}$$

(from Equation (A1.11)). Hence, Equation 2.6d can be set to zero.
The step from $\nabla \cdot \mathbf{j} = 0$ to $\nabla \cdot \mathbf{E} = 0$ is also correct for a layered
Earth giving rise only to horizontal electric fields:

$$\frac{\mathrm{d}\sigma}{\mathrm{d}x} = \frac{\mathrm{d}\sigma}{y} = 0 \quad \text{and} \quad \mathbf{E} = (E_x, E_y, 0). \tag{2.9}$$

For a conductivity distribution that varies in the vertical direction and in one horizontal direction, the divergence of the electric
field parallel to the conductivity boundary is also zero:

$$\frac{\mathrm{d}\sigma}{\mathrm{d}x} = 0 \text{ and } \frac{\mathrm{d}\sigma}{\mathrm{d}y} \neq 0 \quad \text{and} \quad \mathbf{E} = (E_x, 0, 0). \tag{2.10}$$

However, $\nabla \cdot \mathbf{E} \neq 0$ if we consider an electric field perpendicular to a boundary:

$$\frac{d\sigma}{dy} \neq 0 \quad \text{and} \quad \mathbf{E} = (0, E_y, 0). \qquad (2.11)$$

For the case of the Earth, we can consider that a time-varying external magnetic field induces an electric field (according to Faraday's Law (Equation (2.6a)), which in turn induces a secondary, internal magnetic field (according to Ampère's Law (Equation (2.6b)). This process is illustrated in Figure 2.1. By taking the curl of Equation (2.6a) or (2.6b), we can derive a *diffusion equation* in terms of the time-varying electric field, from which information concerning the conductivity structure of the Earth can be extracted. For this purpose we make use of a proven vector identity (Appendix 1; Equation (A1.14)):

$$\nabla \times (\nabla \times \mathbf{F}) = (\nabla \cdot \nabla \cdot \mathbf{F}) - \nabla^2 \mathbf{F} \qquad (2.12)$$

where $\mathbf{F}$ is any vector. For example, taking the curl of Equation (2.6a), substituting Equation (2.6b), and assuming an Earth model for which $\nabla \cdot \mathbf{E} = 0$ yields:

$$\nabla \times (2.6a) \Rightarrow \nabla \times \nabla \times \mathbf{E} = \left(\nabla \cdot \underbrace{\nabla \cdot \mathbf{E}}_{=0}\right) - \nabla^2 \mathbf{E} = -\nabla \times \frac{\partial \mathbf{B}}{\partial t} = \mu_0 \sigma \frac{\partial \mathbf{E}}{\partial t}.$$

$$\therefore \ \nabla^2 \mathbf{E} = \mu_0 \sigma \frac{\partial \mathbf{E}}{\partial t}. \qquad (2.13)$$

Equation (2.13) takes the form of a diffusion equation. Assuming a plane wave with a surface amplitude E_0 and an harmonic time dependence of the form $e^{-i\omega t}$, the right-hand side of Equation (2.13) can be evaluated to give:

$$\nabla^2 \mathbf{E} = i\omega \mu_0 \sigma \mathbf{E}. \qquad (2.14)$$

Similarly,

$$\nabla^2 \mathbf{B} = \mu_0 \sigma \frac{\partial \mathbf{B}}{\partial t} \quad \text{or} \quad \nabla^2 \mathbf{B} = i\omega \mu_0 \sigma \mathbf{B}. \qquad (2.15)$$

In air, $\sigma \to 0$. Therefore, external electromagnetic fields are not significantly attenuated by the air layer between the ionosphere and the Earth's surface.

Equations (2.14) and (2.15) tell us that MT measurements rely on a source of energy that diffuses through the Earth and is exponentially dissipated. Because electromagnetic fields propagate diffusively, MT measurements yield volume soundings (i.e., the response functions are volumetric averages of the sample medium). Gravity measurements also sample volumetrically. However, unlike gravity measurements, which involve a scalar potential, in MT studies we have vector fields. In contrast, seismic techniques are governed by a non-diffusive *wave equation* of the form:

$$\nabla^2 \mathbf{F} = \frac{1}{v^2}\frac{\partial^2 \mathbf{F}}{\partial t^2},\qquad(2.16)$$

where v is the wave propagation velocity.

2.4 The concepts of transfer function and penetration depth

The simplest conceivable geoelectric model is an homogeneous (uniform) half-space, comprised of a zero-conductivity air layer overlying an homogeneous, planar subsurface of conductivity σ_h (Figure (2.2)). We shall use the response of this half-space model to a uniform, time-varying, oscillatory, electromagnetic source field in order to explain the concepts of *transfer function* and *penetration depth*.

Equation (2.13) is a second-order differential equation with a solution of the form (e.g., Boas, 1983):

$$\mathbf{E} = \mathbf{E}_1 e^{i\omega t - qz} + \mathbf{E}_2 e^{i\omega t + qz}.\qquad(2.17)$$

Because the Earth does not generate electromagnetic energy, but only dissipates or absorbs it (Section 2.1; assumption (ii)), arbitrarily large electric field amplitudes cannot be supported within the Earth. This condition implies that $\mathbf{E}_2 = 0$, because $\mathbf{E}$ should diminish as $z \to r_E$ (where r_E is the radius of the Earth).

Taking the second derivative (with respect to depth) of Equation (2.17) we therefore have:

$$\frac{\partial^2 \mathbf{E}}{\partial z^2} = q^2 \mathbf{E}_1 e^{i\omega t - qz} = q^2 \mathbf{E}.\qquad(2.18)$$

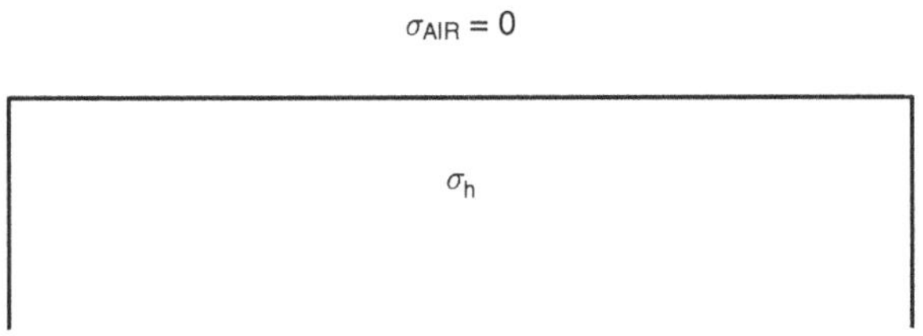

$$\sigma_{AIR} = 0$$

$$\sigma_h$$

Figure 2.2 Uniform half-space composed of a zero-conductivity air layer overlying an homogeneous, planar subsurface of conductivity σ_h.

In our half-space model, $\dfrac{\partial^2 \mathbf{E}}{\partial x^2} = \dfrac{\partial^2 \mathbf{E}}{\partial y^2} = 0$, and we can equate Equation (2.18) with Equation (2.14), yielding an expression for q as:

$$q = \sqrt{i\mu_0\sigma\omega} = \sqrt{i}\sqrt{\mu_0\sigma\omega} = \frac{1+i}{\sqrt{2}}\sqrt{\mu_0\sigma\omega} = \sqrt{\mu_0\sigma\omega/2} + i\sqrt{\mu_0\sigma\omega/2}. \quad (2.19)$$

The inverse of the real part of q:

$$p = 1/\mathrm{Re}(q) = \sqrt{2/\mu_0\sigma\omega} \quad (2.20)$$

is the *electromagnetic skin depth* or *penetration depth* of an electric field with angular frequency ω into a half-space of conductivity σ. The inverse of q:

$$C = 1/q = p/2 - ip/2 \quad (2.21)$$

is referred to as the *Schmucker–Weidelt transfer function* (Weidelt, 1972; Schmucker, 1973). Like p, C depends on frequency and has dimensions of length, but it is complex. For an homogeneous half-space, real and imaginary parts of C have the same magnitudes.

The term *transfer function* invokes an Earth model that describes a linear system with an input and a predictable output. The transfer function C establishes a linear relationship between the physical properties that are measured in the field as follows: from Equation (2.17) with $\mathbf{E}_2 = 0$ we get,

$$(E_x = E_{1x}e^{i\omega t - qz} \Rightarrow) \quad \frac{\partial E_x}{\partial z} = -qE_x \quad (2.22)$$

and, comparing Equation (2.22) to Equation (2.6a) gives

$$\frac{\partial E_x}{\partial z} = -\frac{\partial B_y}{\partial t} = -i\omega B_y = -qE_x. \quad (2.23)$$

Therefore, C can be calculated from measured E_x and B_y fields (or, equivalently E_y and B_x fields) in the frequency domain as:

$$C = \frac{1}{q} = \frac{E_x}{i\omega B_y} = -\frac{E_y}{i\omega B_x}. \quad (2.24)$$

If C is known, then the resistivity of the homogenous half-space can be calculated. Combining Equations (2.19) and (2.24) yields

$$\rho = \frac{1}{\sigma} = \frac{1}{|q|^2}\mu_0\omega = |C|^2\mu_0\omega \quad [\text{Vm A}^{-1}]. \quad (2.25)$$

2.5 Induction in a layered half-space: the concept of apparent resistivity and phase

For an N-layered half-space (Figure 2.3), within every layer n we have a diffusion equation (of the form given in Equation (2.14)) containing conductivity σ_n and a solution (cf. Equation (2.17)) of the form:

$$E_{xn}(q_n, \omega) = E_{1n}e^{i\omega - q_nz} + E_{2n}e^{i\omega + q_nz} = a_n(q_n, \omega)e^{-q_nz} + b_n(q_n, \omega)e^{+q_nz}, \tag{2.26}$$

with q_n being defined similarly to q (in Equation (2.19)), but incorporating the conductivity, σ_n, of the nth layer. In this case, since each layer has limited thickness, $E_{2n} \neq 0$ because z cannot be arbitrarily large.

Similarly to Equation (2.26), the magnetic field within the nth layer is given by:

$$B_{yn}(q_n, \omega) = \frac{q_n}{i\omega}[a_n(q_n, \omega)e^{-q_nz} - b_n(q_n, \omega)e^{+q_nz}]. \tag{2.27}$$

An hypothetical MT sounding penetrating the nth layer could measure E_{xn} and B_{yn}. This would allow the following transfer functions to be computed:

$$C_n(z) = \frac{E_{xn}(z)}{i\omega B_{yn}(z)} \quad \text{and} \quad q_n = \sqrt{i\mu_0\sigma_n\omega}. \tag{2.28}$$

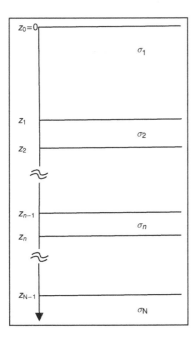

Figure 2.3 N-layered half-space.

By substituting Equations (2.26) and (2.27) into Equation (2.28), we can derive expressions for the transfer functions $C_n(z_{n-1})$ and $C_n(z_n)$ at the top and bottom of the nth layer, respectively.

At the top of the nth layer we have

$$C_n(z_{n-1}) = \frac{a_n e^{-q_n z_{n-1}} + b_n e^{+q_n z_{n-1}}}{q_n(a_n e^{-q_n z_{n-1}} - b_n e^{+q_n z_{n-1}})}. \tag{2.29}$$

and at the bottom of the nth layer

$$C_n(z_n) = \frac{a_n e^{-q_n z_n} + b_n e^{+q_n z_n}}{q_n(a_n e^{-q_n z_n} - b_n e^{+q_n z_n})}. \tag{2.30}$$

Equation (2.30) can be rearranged to yield a term for the ratio a_n/b_n, which we substitute into Equation 2.29 to give:

$$C_n(z_{n-1}) = \frac{1}{q_n} \frac{q_n C_n(z_n) + \tanh[q_n(z_n - z_{n-1})]}{1 + q_n C_n(z_n) \tanh[q_n(z_n - z_{n-1})]}. \tag{2.31}$$

The field components are continuous at the transition from the nth to the $(n-1)$th layer, and it follows that the Schmucker–Weidelt transfer function is also continuous so that

$$C_n(z_n) = \lim_{z \to z_n - 0} C_n(z) = \lim_{z \to z_n + 0} C_{n+1}(z) = C_{n+1}(z_n). \tag{2.32}$$

Inserting the continuity conditions from Equation (2.32) and substituting $l_n = z_n - z_{n-1}$ into Equation (2.31) yields:

$$C_n(z_{n-1}) = \frac{1}{q_n} \frac{q_n C_{n+1}(z_n) + \tanh(q_n l_n)}{1 + q_n C_{n+1}(z_n) \tanh(q_n l_n)}. \tag{2.33}$$

Equation (2.33) is known as Wait's *recursion formula* (Wait, 1954). We can use Wait's recursion formula to calculate the transfer function at the top of the nth layer if the transfer function at the top of the $(n+1)$th layer is known. In order to solve Equation (2.33), we must therefore iterate from the transfer function at the top of the lowermost layer (N), which we define to be an homogeneous half-space, such that (from Equation (2.24))

$$C_N = \frac{1}{q_N}. \tag{2.34}$$

Next, we apply Equation (2.33) ($N-1$ times) until we have the transfer function at the surface of the layered half-space, which can be compared to field data.

We now define *apparent resistivity* as the average resistivity of an equivalent uniform half-space, (the resistivity of which we calculated from the Schmucker-Weidelt transfer function in Section 2.4 (Equation (2.25))):

$$\rho_a(\omega) = |C(\omega)|^2 \mu_0 \omega. \qquad (2.35)$$

Apparent resistivity is one of the most frequently used parameters for displaying MT data. We can also calculate ρ_a from synthetic data by applying the layered-Earth model represented by Equation (2.33) and Figure 2.3. We leave it as an exercise for the reader to show that because ρ_a represents an average taken over the whole volume of the half-space that is penetrated, $\rho_a = 1/\sigma_1$ for penetration depths shallower than z_1 (see Figure 2.3), but that $\rho_a \neq \rho_n$ (where $\rho_n = 1/\sigma_n$) when the penetration depth exceeds the thickness of the first layer.

Because C is complex, we can also extract an *impedance phase*. This is one of the most important MT parameters. The impedance phase, $\phi_{1\text{-D}}$, of our one-dimensional (1-D), layered half-space model can be calculated from

$$\phi_{1\text{-D}} = \tan^{-1}(E_x/B_y). \qquad (2.36)$$

Apparent resistivity and impedance phase are usually plotted as a function of period, $T = 2\pi/\omega$. The functions $\rho_a(T)$ and $\phi(T)$ are not independent of each other, but are linked via the following Kramers–Kroenig relationship (Weidelt, 1972):

$$\phi(\omega) = \frac{\pi}{4} - \frac{\omega}{\pi} \int_0^{\infty} \log\frac{\rho_a(x)}{\rho_0} \frac{\mathrm{d}x}{x^2 - \omega^2}. \qquad (2.37)$$

Equation (2.37) states that the function $\rho_a(T)$ can be predicted from the function $\phi(T)$ except for a scaling coefficient, ρ_0. The fact that in some two-dimensional (2-D) and three-dimensional (3-D) conductivity distributions the form of $\rho_a(T)$ is predictable from the impedance phase, whereas the absolute level is not, reflects the 'distortion' or 'static shift' phenomenon. This is explained in more detail in Chapter 5.

In order to estimate the effect of conductive or resistive layers on the impedance phase, let us consider two idealised two-layer models. Model I consists of an infinitely resistive layer of thickness $h = l_1$, covering an homogeneous half-space with resistivity ρ^*. Model II consists of a thin layer with *conductance* (the product of conductivity and thickness) $\tau = \sigma_1 l_1$, again overlying an homogeneous half-space with resistivity ρ^*. For both models, the penetration depth in the top layer can be considered large compared to the thickness of the top layer, but for different reasons: in model I the resistive layer can be thick, because infinite penetration depths are possible, whereas in model II, the top layer is conductive but assumed to be so thin that the electromagnetic field is only moderately attenuated. Therefore, for both models we have:

$$|q_1 l_1| \ll 1 \quad \text{and, therefore,} \quad \tanh(q_1 l_1) = q_1 l_1.$$

Hence, in both models Equation (2.33) reduces to

$$C_1(z = 0) = -\frac{C_2 + l_1}{1 + q_1 C_2 q_1 l_1}. \tag{2.38}$$

In model I, we have $\sigma_1/\sigma_2 \ll 1$ and therefore $|q_1 C_2| \ll 1$. Therefore,

$$C_1 = C_2 + l_1 = C_2 + h. \tag{2.39}$$

In model II, we have $q_1^2 = i\omega\mu_0\sigma_1$, $l_1 \ll |C_2|$ and $\sigma_1 l_1 = \tau$. Therefore,

$$C_1 = \frac{C_2}{1 + i\omega\mu_0\tau C_2}. \tag{2.40}$$

(The condition for l_1 states that l_1 is so thin that most of the attenuation of the electromagnetic field occurs in the second layer).

In model I – because h is a real number – only the real part of the *complex number* C is enlarged. (i.e., the real part of C_1 becomes larger than the real part of C_2). A comparison of Equations (2.24) and (2.36) tells us that the phase of C and the magnetotelluric phase (Equation (2.36)) are linked according to

$$\phi = \arg C + 90°. \tag{2.41}$$

Given that real and imaginary parts of the transfer function C_2 of the homogeneous half-space are equal in magnitude (Equation (2.21)), the magnetotelluric phase (Equation (2.36)) of the homogeneous half-space is 45°. The magnetotelluric phase that is calculated from C_1 will therefore be greater than 45° for model I. In model II, real and imaginary parts of C_1 are reduced by the attenuation of the electromagnetic field in the conducting top layer, and the magnetotelluric phase, ϕ, is less than 45°. Magnetotelluric phases that are greater than 45° are therefore diagnostic of substrata in which resistivity decreases with depth (e.g., Figure 2.4(a)), and lead to a model from which h can be determined (Equation (2.39)). On the other hand, magnetotelluric phases that are less than 45° are diagnostic of substrata in which resistivity increases with depth (e.g., Figure 2.4(b)), and lead to a model from which τ can be determined (Equation (2.40).

A convenient display of data that can be plotted on the same $\rho(z)$ graph as a 1-D model is provided by the ρ^*-z^* transform (Schmucker, 1987), where z^* is defined by $\text{Re}(C)$, and ρ^* is given by

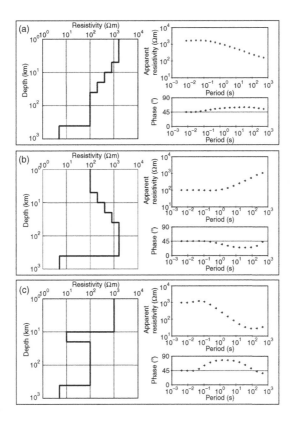

Figure 2.4 (a)–(c)
Period-dependent apparent
resistivities and impedance
phases generated by a layered
half-space model in which: (a)
resistivity decreases with
depth. At the shortest periods
the impedance phases are 45°,
consistent with a uniform
half-space model and increase
above 45° at ~10 Hz, consistent
with the decrease in resistivity;
(b) resistivity increases with
depth. At the shortest periods
the impedance phases are 45°,
consistent with a uniform
half-space model, and decrease
below 45° at ~10 Hz, consistent
with the increase in resistivity;
(c) is incorporated a 10-km-
thick high-conductivity layer
representing the mid crust. The
three models shown in (a), (b)
and (c) all terminate with a
5 Ωm half-space at 410 km.

$$\rho^* = 2\rho_a \cos^2\phi \qquad \text{for } \phi > 45° \text{ (Model I)} \tag{2.42}$$

$$\rho^* = \rho_a/(2\sin^2\phi) \quad \text{for } \phi < 45° \text{ (Model II).} \tag{2.43}$$

If $\phi = 45°$, Equations (2.42) and (2.43) both yield $\rho^* = \rho_a$, and
the model reverts to the one of an homogeneous half-space (i.e., in
the absence of the resistive/conductive layer, $h = 0/\tau = 0$).

To describe the transfer function associated with a three-layer
'sandwich' model consisting of a conductive middle layer with conductance τ and a resistive upper layer of thickness h, Equations
(2.39) and (2.40) can be combined to yield

$$C_1 = C_2 + h = \frac{C_3 + h}{1 + i\omega\mu_0\tau C_3}, \tag{2.44}$$

where C_1, C_2 and C_3 are the transfer functions at the top of the
uppermost, middle and third layer, respectively. Typical forms of the
apparent resistivities and impedance phases as a function of period
generated by such a 'sandwich' model are depicted in Figure 2.4(c).

Figure 2.5 Simple 2-D model composed of quarter-spaces with different conductivities meeting at a vertical contact (planar boundary extending to infinity – i.e., striking – in the x direction). Conservation of current across the contact, where the conductivity changes from σ_1 to σ_2, leads to the y-component of the electric field, E_y, being discontinuous. For this idealised 2-D case, electromagnetic fields can be decoupled into two independent modes: one incorporating electric fields parallel to strike with induced magnetic fields perpendicular to strike and in the vertical plane (**E-polarisation**); the other incorporating magnetic fields parallel to strike with induced electric fields perpendicular to strike and in the vertical plane (**B-polarisation**).

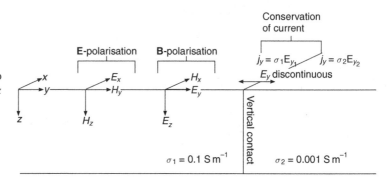

2.6 Induction at a discontinuity: a simple two-dimensional (2-D) world and the concept of E- and B-polarisation

The physical principle governing induction at a discontinuity is conservation of current. Figure 2.5 shows a very simple 2-D scenario with a vertical contact between two zones of different conductivity, σ_1 and σ_2. The current density, (j_y), across the boundary is given by:

$$j_y = \sigma E_y. \tag{2.45}$$

Since current must be conserved across the boundary, the change in conductivity demands that the electric field, E_y, must also be discontinuous. All other components of the electromagnetic field are continuous across the boundary.

The scenario shown in Figure 2.5 may represent a dyke or a fault with an approximately constant conductivity along its strike. For a body with infinite along-strike extension, or one with an along-strike wavelength significantly longer than the penetration depth, there are no along-strike (x-direction in Figure 2.5) field variations (i.e., $\frac{\partial}{\partial x} = 0$) and Equations (2.6a) and (2.6b) can be expanded as:

$$\frac{\partial(E_x - E_y + E_z)}{\partial y} + \frac{\partial(E_x - E_y + E_z)}{\partial z} = \mathrm{i}\omega(B_x - B_y + B_z) \tag{2.46a}$$

$$\frac{\partial(B_x - B_y + B_z)}{\partial y} + \frac{\partial(B_x - B_y + B_z)}{\partial z} = \mu_0\sigma(E_x - E_y + E_z). \tag{2.46b}$$

Furthermore, for the ideal 2-D case, electric and magnetic fields are mutually orthogonal: an electric field parallel to strike induces magnetic fields only perpendicular to strike and in the vertical plane, whilst a magnetic field parallel to strike induces electric fields only perpendicular to strike and in the vertical plane (Figure 2.5). Therefore, Equations (2.46a) and (2.46b) can be decoupled into two independent modes: one incorporating electric fields parallel to strike (**E**)-polarisation, the other incorporating magnetic fields parallel to strike (**B**-polarisation).

The **E**-polarisation (sometimes referred to as the *transverse electric*, or *TE mode*) describes currents flowing parallel to strike (x-direction in Figure 2.5) in terms of the electromagnetic field components E_x, B_y and B_z:

$$\left.\begin{array}{l} \dfrac{\partial E_x}{\partial y} = \dfrac{\partial B_z}{\partial t} = i\omega B_z \\[3mm] \dfrac{\partial E_x}{\partial z} = \dfrac{\partial B_y}{\partial t} = -i\omega B_y \\[3mm] \dfrac{\partial B_z}{\partial y} - \dfrac{\partial B_y}{\partial z} = \mu\sigma E_x \end{array}\right\} \textbf{E}\text{-polarisation.} \qquad (2.47a)$$

The **B**-polarisation (sometimes referred to as the *transverse magnetic* or *TM mode*) describes currents flowing perpendicular to strike (y-direction in Figure 2.5) in terms of the electromagnetic field components B_x, E_y and E_z:

$$\left.\begin{array}{l} \dfrac{\partial B_x}{\partial y} = \mu_0\sigma E_z \\[3mm] \dfrac{-\partial B_x}{\partial z} = \mu_0\sigma E_y \\[3mm] \dfrac{\partial E_z}{\partial y} - \dfrac{\partial E_y}{\partial z} = i\omega B_x \end{array}\right\} \textbf{B}\text{-polarisation.} \qquad (2.47b)$$

Since E_y is discontinuous across a vertical contact, the impedances – Z_{yx} (the ratio $\mu_0 E_y/B_x$) and Z_{yy} (the ratio $\mu_0 E_y/B_y$) – associated with E_y are also discontinuous. However, for the simple 2-D case shown in Figure 2.5, Z_{yy} is zero, and we need consider only Z_{yx}. From Equation (2.45), the magnitude of the discontinuity in E_y and therefore in Z_{yx} is σ_2/σ_1. Therefore, from Equation (2.35), there will be a discontinuity in the apparent resistivity, ρ_{yx}, perpendicular to strike of magnitude $(\sigma_2/\sigma_1)^2$.

Figure 2.6 shows ρ_{yx} values for a range of frequencies as a function of distance from the vertical contact shown in Figure 2.5. As a consequence of the discontinuous behaviour exhibited by ρ_{yx}, **B**-polarisation resistivities tend to resolve lateral conductivity variations better than **E**-polarisation resistivities. However, the

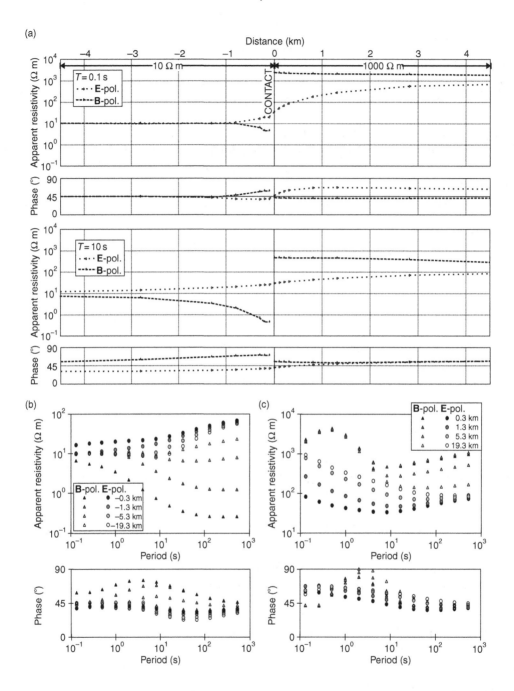

Figure 2.6 (a) Comparison of **E**- and **B**-polarisation apparent resistivities and impedance phases as a function of distance from the vertical contact shown in Figure 2.5 for periods of 0.1 and 10 s. Whereas the **E**-polarisation apparent resistivities vary smoothly across the contact, the **B**-polarisation apparent resistivities are discontinuous. (b) Apparent resistivities and impedance phases as a function of period at distances of −0.3, −1.3, −5.3 and −19.3 km (over the 10 Ω m quarter-space) from the vertical contact. (c) Apparent resistivities and impedance phases as a function of period at distances of +0.3, +1.3, +5.3 and +19.3 km (over the 1000 Ω m quarter-space).

E-polarisation has an associated vertical magnetic field. Vertical magnetic fields are generated by lateral conductivity gradients and boundaries, and spatial variations of the ratio H_z/H_y can be used to diagnose lateral conductivity contrasts from the E-polarisation (Section 2.8).

2.7 Adjustment length

So far, we have stressed the fact that MT measurements yield a volume sounding, but visualising a hemispherical volume is far too simplistic, because the conductivity structure of the Earth varies laterally as well as with depth. Lateral conductivity variations distort the sounding volume away from that of an idealised hemisphere, because electromagnetic fields of a given sounding period penetrate less deeply into high-conductivity heterogeneities (Figure 2.7). Therefore, we need to consider not only the penetration depth of data, but also the *horizontal adjustment length*, which is the lateral distance to which an MT transfer function of a given period is sensitive. Penetration depths and horizontal adjustment lengths are not equivalent. In fact, conductivity anomalies laterally displaced by 2–3 times the penetration depth may effect the MT transfer functions. This is demonstrated in Figure 2.8, where a conductor located in a layered half-space 30 km away from site A begins to affect the MT transfer functions (particularly the impedance phases) at periods of $\sim$10 s. Substituting the apparent resistivity (100 Ω m) at 10 s into Equation (1.1) or Equation (2.20) yields a penetration depth of $\sim$16 km. In this model, fields with periods of 10 s therefore have an horizontal adjustment length that is approximately twice their penetration depth. In general, perturbations in current levels owing to a surficial conductivity heterogeneity overlying a resistive layer are attenuated to e^{-1} of their amplitudes at the

Figure 2.7 Simplistic depiction of the propagation of electromagnetic wave fronts in a half-space of conductivity σ_1 within which a conductive block of conductivity σ_2 is embedded. Electromagnetic fields of a given sounding period penetrate less deeply into the body with higher conductivity, thus distorting the sounding volume away from that of an idealised hemisphere.

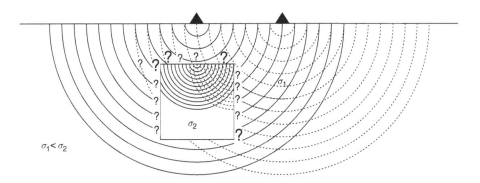

Figure 2.8 Apparent resistivity
and impedance phase curves
for a site A installed 30 km from
a 1-km-deep ocean. The
impedance phases diverge
from the response expected for
a 100 Ω m half-space (directly
underlying site A) at periods
of ~10 s. For example, at ~8 s,
the phase splitting (Δϕ) is ~2°.
Therefore the horizontal
adjustment length at ~10 s is
approximately twice the
penetration depth ($p \approx 14$ km).

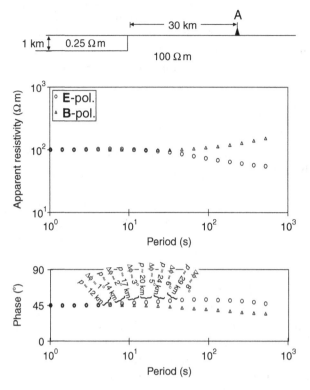

boundary of the heterogeneity for a horizontal adjustment distance
given by:

$$\sqrt{(\sigma \Delta z_1)(\rho \Delta z_2)}, \qquad (2.48)$$

where $(\sigma \Delta z_1)$ is the conductivity × thickness product of the surface
layer and $(\rho \Delta z_2)$ is the resistivity × thickness product of the subsurface layer (Ranganayaki and Madden, 1980).

　　The conductivity anomaly depicted in Figure 2.8 might represent a shallow ocean. An example of how failure to take into
account the horizontal adjustment distance in the presence of coast-
lines can lead to artefacts when modelling measured data is given in
Section 7.3. As a consequence of the horizontal adjustment length,
continental MT measurements made in coastal regions can be
highly sensitive to oceanic mantle conductivities (Mackie *et al.*,
1988).

2.8 Induction arrows

Induction arrows are vector representations of the complex
(i.e., containing real and imaginary parts) ratios of vertical to

horizontal magnetic field components. Since vertical magnetic fields are generated by lateral conductivity gradients, induction arrows can be used to infer the presence, or absence of lateral variations in conductivity. In the *Parkinson convention*, which has been adopted as the standard within the MT community, the vectors point **towards** anomalous internal concentrations of current (Parkinson, 1959). Occasionally, the *Wiese convention* (Wiese, 1962) is used, in which case the vectors point **away** from internal current concentrations. The vectors are sometimes called *tipper vectors*, because they transform or tip horizontal magnetic fields into the vertical plane according to the relationship:

$$H_z(\omega) = (\, T_x(\omega) \quad T_y(\omega)\,) \begin{pmatrix} B_x/\mu_0 \\ B_y/\mu_0 \end{pmatrix}. \tag{2.49}$$

In a 2-D Earth, induction arrows are associated only with the E-polarisation (compare Equations (2.47a) and (2.47b) in Section 2.6). Thus, insulator–conductor boundaries extending through a 2-D Earth give rise to induction arrows that orientate perpendicular to them, and have magnitudes that are proportional to the intensities of anomalous current concentrations (Jones and Price, 1970), which are in turn determined by the magnitude of the conductivity gradient or discontinuity. However, an absence of induction arrows at a single site does not necessarily confirm an absence of laterally

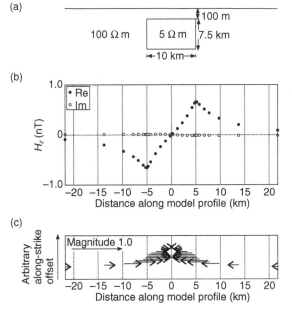

Figure 2.9 (a) 2-D model incorporating a 10 km × 7.5 km, 5 Ω m conductive rod of infinite length embedded at 100 m depth in a 100 Ω m half-space. (b) Form of the vertical magnetic field traversing the 2-D conductivity anomaly shown in (a). (c) Parkinson induction arrows along a profile of sites traversing the conductivity anomaly shown in (a).

displaced conductivity boundaries. Figure 2.9(b) shows the form of the vertical magnetic field traversing the 2-D conductivity anomaly shown in Figure 2.9(a). Notice that at the centre of the idealised anomaly, the vertical magnetic field decays to zero, and then reverses its sense. The changes in magnitude and orientation of the induction arrows along a profile of sites traversing the conductivity anomaly is shown in Figure 2.9(c).

2.9 The impedance tensor and a preview of three-dimensionality

The MT technique is a passive technique that involves measuring fluctuations in the natural electric (**E**) and magnetic (**B**) fields in orthogonal directions at the surface of the Earth. The orthogonal components of the horizontal electric and magnetic fields are related via a complex *impedance tensor*, $\underline{\underline{Z}}$:

$$\begin{pmatrix} E_x \\ E_y \end{pmatrix} = \begin{pmatrix} Z_{xx} & Z_{xy} \\ Z_{yx} & Z_{yy} \end{pmatrix} \begin{pmatrix} B_x/\mu_0 \\ B_y/\mu_0 \end{pmatrix} \quad \text{or} \quad \mathbf{E} = \underline{\underline{Z}}\mathbf{B}/\mu_0 \qquad (2.50)$$

$\underline{\underline{Z}}$ is complex, being composed of both real and imaginary parts. Therefore, each component, Z_{ij}, of $\underline{\underline{Z}}$ has not only a magnitude, but also a phase (cf. Equations (2.35) and (2.36)):

$$\rho_{a,ij(\omega)} = \frac{1}{\mu_0\omega}\left|Z_{ij}(\omega)\right|^2 \qquad (2.51)$$

$$\phi_{ij} = \tan^{-1}\left(\frac{\mathrm{Im}\{Z_{ij}\}}{\mathrm{Re}\{Z_{ij}\}}\right). \qquad (2.52)$$

To illustrate the physical meaning of *impedance phase*, we will use an analogy based on thermal conditions within a wine cellar (Figure 2.10). The temperature inside the wine cellar doesn't vary as much as the temperature outside, because some of the signal gets attenuated as it penetrates through the walls, which have a lower thermal conductivity than air. This attenuation is analogous to the

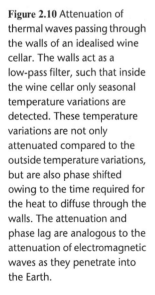

Figure 2.10 Attenuation of thermal waves passing through the walls of an idealised wine cellar. The walls act as a low-pass filter, such that inside the wine cellar only seasonal temperature variations are detected. These temperature variations are not only attenuated compared to the outside temperature variations, but are also phase shifted owing to the time required for the heat to diffuse through the walls. The attenuation and phase lag are analogous to the attenuation of electromagnetic waves as they penetrate into the Earth.

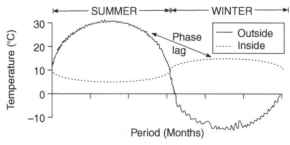

attenuation of electromagnetic waves as they penetrate into the Earth. In addition to attenuation, there is also a phase lag, owing to the time required for the heat to diffuse through the walls – (i.e., the time at which the outside air temperature reaches a maximum will not be the time at which the inside temperature peaks, which will be later). In the ideal case, a stone-walled cellar might actually be colder in the summer than in the winter – a gourmet's delight: a nice cool white wine to accompany a seafood salad in summer, and a room temperature red wine to enjoy with a hearty roast in winter!

Being a tensor, $\underline{\underline{Z}}$ also contains information about dimensionality and direction. For a 1-D Earth, wherein conductivity varies only with depth, the diagonal elements of the impedance tensor, Z_{xx} and Z_{yy} (which couple parallel electric and magnetic field components) are zero, whilst the off-diagonal components (which couple orthogonal electric and magnetic field components) are equal in magnitude, but have opposite signs, i.e.,:

$$\left. \begin{array}{l} Z_{xx} = Z_{yy} = 0 \\ Z_{xy} = -Z_{yx} \end{array} \right\} \text{1-D.} \tag{2.53}$$

For a 2-D Earth, in which conductivity varies along one horizontal direction as well as with depth, Z_{xx} and Z_{yy} are equal in magnitude, but have opposite sign, whilst Z_{xy} and Z_{yx} differ, i.e.,:

$$\left. \begin{array}{l} Z_{xx} = -Z_{yy} \\ Z_{xy} \neq -Z_{yx} \end{array} \right\} \text{2-D.} \tag{2.54}$$

For a 2-D Earth with the x- or y-direction aligned along *electromagnetic strike*, Z_{xx} and Z_{yy} are again zero. Mathememematically, a 1-D anisotropic Earth is equivalent to a 2-D Earth.

With measured data, it is often not possible to find a direction in which the condition that $Z_{xx} = Z_{yy} = 0$ is satisfied. This may be due

Figure 2.11 Scale dependence of dimensionality. A 3-D body of conductivity σ_2 embedded in a conductivity half-space of conductivity σ_1 induces a 1-D responses for MT sounding periods that are sufficiently short that their skin depths are small compared to the shortest dimensions of the 3-D body. As the sounding period increases, the inductive scale length will eventually extend sufficiently to encompass at least one edge of the anomaly, and the MT transfer functions appear multi-dimensional. For sufficiently long periods, such that the electromagnetic skin depth is very much greater than the dimensions of the anomaly, the inductive response of the anomaly becomes weak, but a so-called galvanic response that is frequency independent (i.e., real) remains (see Chapter 5).

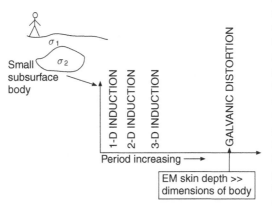

to distortion or to 3-D induction (or both). Generally, the dimensionality evinced by data is scale dependent. Consider any generalised, homogeneous, 3-D conductivity anomaly embedded in an otherwise uniform Earth. For short MT sounding periods, corresponding to electromagnetic skin depths that are small compared to the shortest dimensions of the anomaly, the transfer functions should appear 1-D. As the sounding period increases, the *inductive scale length* will eventually extend sufficiently to encompass at least one edge of the anomaly, and the transfer functions appear 2-D. As the sounding period increases further, edge effects from all sides of the anomaly will eventually be imposed on the transfer functions, resulting in transfer functions that are evidently 3-D (Figure 2.11). For sufficiently long periods, such that the electromagnetic skin depth is very much greater than the dimensions of the anomaly, the inductive response of the anomaly becomes weak, but a non-inductive response persists. The non-inductive response of the anomaly creates a frequency-independent distortion of MT transfer functions that can be stripped away (as we shall see in Chapter 5).

Chapter 3
Planning a field campaign

The choice of equipment used in a particular survey should depend on the depth range under consideration: in crustal studies, induction coil magnetometers are used frequently, the sampling is quick and the 'processing' (described in Chapter 4) is usually performed in the field. Fluxgate magnetometers provide a response at longer periods than induction coils, and are used if larger penetration depths are under consideration. In many cases, data from very short to very long periods are desirable, and two different sensors are combined at each site. It is vital that anybody writing or modifying processing software has access to all information regarding the analogue electronics of the system (e.g., calibration coefficients for filters) that is to be used in conjunction with the software.

We suggest a rule of site spacing: not too close and not too sparse. The question whether we should deploy magnetotelluric sites along a profile, or as a 2-D array is discussed in the context of the geological complexity of the target area, the available hardware and the financial resources. In many cases, a trade-off has to be found between the desire to have many sites and hence a good spatial resolution and the wish to achieve high-quality data by occupying sites for a long time.

3.1 Target depths and choosing the right sensors and equipment

3.1.1 Considering the period range

From the definition of *penetration depth* (Equation (2.20)), we can estimate a period range associated with a particular depth range of interest, provided that an estimate of the subterranean conductivity is known. Of course, the conductivity is not known exactly until we have made some measurements, but, for the purpose of planning a field campaign, we can assume a substitute medium with average *bulk conductivities* of $0.001 \, \mathrm{S \, m^{-1}}$, $0.02 \, \mathrm{S \, m^{-1}}$ and $0.1 \, \mathrm{S \, m^{-1}}$ for Palaeozoic, Mesozoic and tertiary crust, respectively, and $0.02 \, \mathrm{S \, m^{-1}}$ for the continental upper mantle. (The meaning of 'bulk conductivity' will be explored further in Chapter 8). Note that the substitute medium is assumed only for the purpose of ascertaining the appropriate sensors to deploy, and will be superseded by a more complex model once data have been acquired.

From the inverse of Equation (2.20):

$$T = \mu_0 \sigma \pi p^2 \tag{3.1}$$

we find, for example, that periods spanning $\sim$0.002 s – $\sim$10 s would be required in order to explore crustal depths of order 1–50 km within an old, resistive craton. If, on the other hand, we suppose that a resistive ($1000 \, \Omega \, \mathrm{m}$) upper crust is underlain by a conductive mid crust (having conductance $\tau = 1000 \, \mathrm{S}$) at $h = 20 \, \mathrm{km}$ depth, and a lower crust of resistivity $1000 \, \Omega \, \mathrm{m}$, we can calculate an approximate *Schmucker–Weidelt transfer function*:

$$C = \frac{C_3}{1 + i\omega\mu_0\tau C_3} + h, \tag{3.2}$$

where C_3 is the transfer function of the lower crustal homogeneous *half-space* that is assumed to terminate the model. The derivation can be found in Section 2.5. For the three-layer scenario outlined, $C_3 = 50 \, \mathrm{km} - 50\mathrm{i} \, \mathrm{km}$ at $T = 10 \, \mathrm{s}$. The imaginary term in the denominator of the first term on the right-hand side of Equation (3.2) describes the attenuation owing to the mid-crustal conductor in our three-layer model. At $T = 10 \, \mathrm{s}$, $|i\omega\mu_0\tau C_3| \approx 15$, and because this term is inversely proportional to T (i.e., $|i\omega\mu_0\tau C_3| \propto 1/T$), periods 150 times longer than 10 s (i.e., 1500 s) are required for the attenuation to be less than 0.1. Thus, we might require sounding periods spanning 0.002 s to 1500 s in order to explore depths in the range 1–50 km within a thick cratonic crust containing a moderately conductive mid-crustal layer.

For mantle studies, we require periods in the range 10–$10^5 \, \mathrm{s}$, in order to provide short-period control over the nature of the overlying crust and long-period control on the *transition zones* at 410–660 km. The significance of the transition zones for MT studies is explained in Chapter 5 (Section 5.8) and Chapter 8 (Section 8.2).

3.1.2 Magnetic sensors

Two principal types of magnetic sensors are used in MT studies: *induction coils* and *fluxgate magnetometers*. Induction coils usually consist of a coil of copper wire wound onto a high-permeability core, sealed within a shock-resistant casing. A set of three induction coils plus a spirit level and a compass for aligning their axes are required in order to measure all three components of the time-varying magnetic field. The output voltage of an induction coil is directly proportional to the number of loops in the coil and their cross-sectional area (as explained in any textbook on general or electromagnetic physics, e.g., Tipler, 1991). Therefore, the design of coils suitable for MT fieldwork is essentially a compromise between transportability (i.e., weight and length), and *sensitivity*. Because the response of an induction coil is governed by the rate of change of magnetic flux within the coil, which is directly proportional to $d\mathbf{B}/dt$, the sensitivity of induction coils is highest for the case of rapidly varying (i.e., short-period) fields. Fluxgate magnetometers generally consist of three ring-core sensors (elements composed of two cores of easily saturable, high-permeability material, oppositely wound with coaxial excitation coils) mounted onto a plate such that their axes are mutually orthogonal, and enclosed in a waterproof capsule that can be buried in the Earth (Figure 3.1). The construction usually includes a spirit level for ensuring that the plate is planted horizontally in the ground and a screw mechanism that allows fine tuning of the orientation. Fluxgate magnetometers rely on the principle of hysteresis, (which occurs when the core of the sensor is driven to saturation by an alternating current in the surrounding coil). Hysteresis cycles generate an output that is sensitive to the intensity of the time-varying exciting magnetic field. Therefore fluxgate magnetometers are suitable for measuring

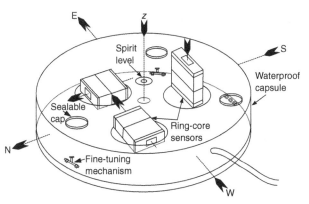

Figure 3.1 Fluxgate magnetometer consisting of three mutually perpendicular ring-core sensors enclosed in a waterproof capsule. Sealable caps provide access to screw threads that allow the user to fine-tune the orientation of the plate onto which the ring-core sensors are mounted, and a built-in spirit level indicates when the instrument is level (i.e., horizontal).

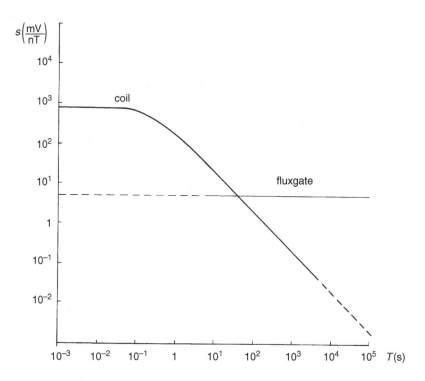

Figure 3.2 Typical period-dependent sensitivities, S, of an induction coil magnetometer and of a fluxgate magnetometer. Dashed lines indicate period ranges where a particular sensor is not used.

long-period magnetic field variations, which have high amplitudes (see Figure 1.1); for periods shorter than a threshold period, the amplitude of the natural signal becomes weaker than the noise of the sensor. More details about fluxgate magnetometers can be found in the review by Primdahl (1979).

Figure 3.2 summarises the response characteristics of induction coils versus those of fluxgate magnetometers. Induction coil magnetometers can respond well to magnetic fluctuations with periods between 0.001 s to 3600 s, whereas fluxgate magnetometers cover periods ranging from 10 s to 100 000 s. Therefore, users who want to cover the entire period range from 0.001 s to 100 000 s will combine the use of induction coils and fluxgate magnetometers.

The $2\frac{1}{2}$ decades of overlap between the data produced by the two types of magnetometer can be useful for checking consistency of results, particularly during instrument development. Alternatively, a high-quality induction coil magnetometer can be combined with a cheaper fluxgate with higher noise level. In this case, the threshold period of the fluxgate is shifted towards longer periods, and the overlap band (Figure 3.2) is smaller.

3.1.3 Electric field sensors

Electric field fluctuations are determined by measuring the potential difference, U, between pairs of electrodes, which are connected via a shielded cable to form a dipole and buried in the ground at known distances, d, 10–100 m apart:

$$E = \frac{U}{d}. \tag{3.3}$$

Two dipoles are required in order to ascertain the two horizontal components of the electric field. These dipoles are typically configured orthogonal to each other, with one dipole oriented in the magnetic north–south (N–S) direction, and the other in the magnetic east–west (E–W) direction. Steel nails can suffice as electrodes for high-frequency audiomagnetotelluric (AMT) measurements, but longer-period measurements require non-polarisable electrodes in which electrochemical effects (which modify the potential difference that is registered) are avoided as far as possible. Non-polarisable electrodes usually consist of a porous pot containing a metal (e.g., silver [Ag]) in contact with a salt of the same metal (e.g., silver chloride [AgCl]). Petiau and Dupis (1980) provide an overview and comparison of different electrode types.

Junge (1990) adapted the Ag–AgCl ocean-bottom MT electrode of Filloux (1973, 1987) for long-period land measurements. In this design, the oceanic environment is simulated by a saturated KCl solution, and the electrical contact between the KCl solution and the ground is provided by a ceramic diaphragm (Figure 3.3). This design allows for MT measurements in the period range of the daily variation (e.g., Simpson, 2001b).

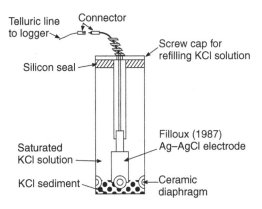

Telluric line to logger

Connector

Screw cap for refilling KCl solution

Silicon seal

Saturated KCl solution

KCl sediment

Filloux (1987) Ag–AgCl electrode

Ceramic diaphragm

Figure 3.3 Cross-section through a cylindrical silver–silver chloride (Ag–AgCl) electrode of the type commonly used in long-period magnetotelluric measurements.

During data recording, it is very important that electrodes are not exposed to temperature variations spanning the period range of interest. In the most demanding case, where electric daily variations are measured, the top end of an electrode should be buried at least 50 cm below the surface. The penetration of a thermal wave into the ground is governed by a *diffusion equation* (cf. Equation (2.14)), where the electrical diffusivity $1/\mu_0\sigma$ is replaced by the thermal diffusivity (e.g., Stacey, 1992). A depth of 50 cm is more than twice the penetration depth of a thermal wave with period of 1 day (assuming a thermal diffusivity of $10^{-6}\,\mathrm{m^2\,s^{-1}}$), and the amplitude of the daily temperature variation is therefore attenuated to e^{-2} of its surface value.

Some MT practitioners advocate the use of wet bentonite (clay) in electrode holes to facilitate a better contact between the electrodes and the ground. This technique is not recommended for long-period MT measurements, because the bentonite dries out over the duration of the recording, causing the potential difference to drift.

3.1.4 Data-acquisition systems

There are many different data-acquisition systems or 'dataloggers' available in geophysics, a number of which have been particularly designed with electromagnetic induction studies in mind. In this section, we outline the main points that potential users should consider when buying or designing a datalogger. Key considerations are the rate at which the datalogger should sample data, signal resolution and the type and size of the medium used for data storage.

In order to determine the rate at which the electromagnetic *time series* needs to be sampled, we need to understand a fundamental tenet associated with sampling processes. This is described in detail in text books on digital time series processing (e.g., Otnes and Enochson, 1972) and is only summarised here. The *sampling theorem* states that if a time series is sampled at Δt intervals, the digital time series adequately describes signals with periods longer than $2\Delta t$ (which is known as the *Nyquist period, T_{NY}*), whereas periods shorter than $2\Delta t$ are undersampled, and generate an artificial, low-frequency signal in the digital time series. Distortion of digital time series by undersampling is known as *aliasing* and has severe consequences for the design of dataloggers. A simple example of aliasing is shown in Figure 3.4, in which a sinusoidal signal is sampled at regular time intervals that are longer than the half-period of the original signal. Because the sampling is too sparse, the original signal

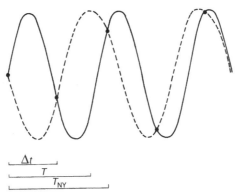

Figure 3.4 Example of *aliasing* in the time domain. The sampling interval (Δt) is longer than half the period (T) of the original signal (solid line). Therefore we cannot retrieve the original (analogue) signal after digitisation, but instead infer a signal (dashed line) with a longer period (T_{NY}) from the digital data (discrete points).

cannot be retrieved from the sampled data. Instead, the reconstructed signal has a longer wavelength than the original signal. If T_0 is the shortest evaluation period of interest, and Δt the sample rate, we therefore have the requirement that $\Delta t \leq T_0/2$. In practice, $\Delta t = T_0/4$ is used. If the data storage medium is cheap and of high capacity, a shorter time increment may also be used in conjunction with analogue filters (see later).

In order to ascertain whether a 16-bit or a 24-bit analogue-to-digital (A/D) converter is required, we next consider sample resolution. If we use a 16-bit A/D converter (Figure 3.5), the ratio of the largest amplitude which the system can handle to the smallest variation it can resolve is $2^{16} = 65\,536$. For the case of a long-period MT system, the largest magnetic variations are created by *magnetic storms*, and are not expected to exceed $\pm\,500\,$nT. The amplifier for the fluxgate magnetometer should therefore be designed in such a way that 65 536 is identified with $1000\,$nT, so that the *least count* '1' is identified with $1000\,$nT/65 536 = 15.2588 pT (picotesla). This least count is the smallest magnetic change that a data-acquisition system with a 16-bit A/D converter can resolve. Whether there is anything to gain by using a 24-bit A/D converter instead of a 16-bit A/D

Figure 3.5 Example of the 16-bit A/D converter that can resolve 15 pT, and handle a maximal variational field strength of $\pm500\,$nT (the 16th bit transports the sign, because positive and negative signals occur). The 24-bit A/D converter has a theoretical resolution of 6 pT, if the geomagnetic main field, with a strength of the order $\pm50\,000\,$nT, is recorded (rather than compensated).

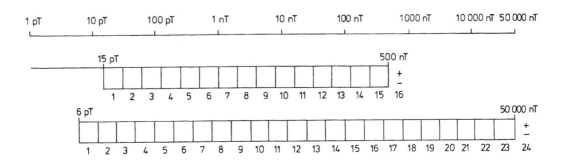

converter depends on the noise level within the fluxgate magnetometer. If the noise of the fluxgate is significantly lower than 15 pT, then a 24-bit A/D converter could help to shift the resolution threshold of the system towards lower-intensity magnetic variations. Otherwise, the increased sensitivity of the system to natural signals will be masked by noise from the fluxgate magnetometer.

In determining the sample resolution, we considered only variations in the magnetic field. This assumes that we have compensated the geomagnetic main field (of order 50 000 nT) of the Earth prior to amplification. This *compensation* (sometimes called *backing-off* the geomagnetic main field) is realised by establishing a stable voltage – equivalent to the voltage owing to the geomagnetic main field acting on the sensor – across an analogue feedback circuit, and subtracting this voltage from the output of the sensor prior to amplification. Self-potentials with amplitudes larger than the amplitude of electric field variations are compensated in a similar way. Maintaining stability of the compensation voltage is crucial: a temperature effect that changes the resistivities of the resistors within the analogue circuit by only 0.01%, will create a virtual magnetic field of 50 000 nT/10 000 = 5 nT, which is significant compared to the amplitude of the natural magnetic variations.

When digital components were expensive, and dataloggers relied on 16-bit A/D converters, compensation and the associated precise analogue electronics had to be incorporated into the data-acquisition system. However, in the digital age, in which precise analogue electronic circuits are relatively expensive to manufacture compared to A/D converters, modern dataloggers may substitute a 24-bit A/D converter for a 16-bit A/D converter, so that the geomagnetic main field can be recorded rather than compensated. In this case, we identify 100 000 nT with 2^{24}, and the least count is identified with 6 pT (Figure 3.5). (If the 24-bit A/D converter is 'noisy', then it might be better not to use the last 2 bits, in which case we could realistically obtain a least count that is $2^2 \times 6 = 24$ pT). In ocean-bottom studies, recording the components of the main field can also be used to ascertain the otherwise unknown orientation of the sensor.

When deploying a broad-band induction coil magnetometer that provides data in the *period band* 1000 s – 1/1000 s, the factor 10^6 between the smallest and largest amplitudes occurring in this period band (see Figure 1.1) might suggest the need for a high-resolution A/D converter. However, the $\mathbf{B}(T)$ characteristic is exactly cancelled out by the $\delta\mathbf{B}(t)/\mathrm{d}t$ response of the coil (Figure 3.2).

Naturally-induced electric signals should be linear functions of **B**:

$$\mathbf{E} = \frac{i}{q}\omega\mathbf{B} \approx \sqrt{\omega}\mathbf{B} \tag{3.4}$$

and have a smaller period dependence then the magnetic field, adding no additional requirements to the A/D resolution. However, for audiomagnetotelluric acquisition systems, a 24-bit A/D converter can be advantageous in the presence of sources of electrical noise (such as those associated with power lines).

To avoid the aliasing problem mentioned in the introduction to this section, we have to remove all periods shorter than $2\Delta t$ from the analogue signals prior to digitisation. Removal of periods shorter than the Nyquist period is achieved using a low-pass filter[4], which is a sequence of resistor–capacitor (RC) circuits having a cut-off period, T_c:

$$T_c = 2\pi RK, \tag{3.5}$$

where R is resistance and K is capacitance. In analogue filter design, the cut-off frequency is normally defined as the -3 dB point (where the input signal is reduced to $1/\sqrt{2}$ of the maximum value). Given that at the cut-off period the complex ratio of input to output voltage is reduced to $1/\sqrt{2}$, how many circuits are necessary to remove the energy at the Nyquist period? There are two different approaches to this problem (Figure 3.6). In the past, when digital storage media were expensive and precise analogue electronics were readily available, six to eight RC circuits would typically be used. The resulting filter is steep, the cut-off period, T_c, is close to the first evaluation period T_0, and the capacity of the digital storage medium can be reasonably small because a sampling rate of $\Delta t = T_0/4$ is adequate. However, the filters have to be calibrated in the frequency domain, and the *calibration* coefficients have to be taken into account during processing of the MT data. Therefore, we need to know the calibration coefficients of individual filters, and we need to know which filter was used in a particular channel of the datalogger. A possible simplification would be to give all channels – electric and magnetic – the same filter design, and hope that we don't need to know the exact design if we are only interested in the ratio E_{ij}/B_{ij}. But *are* they all the same? In the vicinity of the cut-off period, a 10%

---

[4] In ocean-bottom studies, it is not necessary to incorporate a low-pass filter into the data-acquisition system, because the ocean acts as a natural low-pass filter, as explained in Section 9.4.

Figure 3.6 Filter response, w,
versus period for two long-
period data-acquisition
systems both having $T_c = 10$ s.
Solid curve represents the
response of the steep,
analogue, anti-alias filters
typically used in an old system
with sampling rate, Δt_{old}, and
Nyquist period, $T_{NY, old}$.
Dashed line represents the
response characteristics typical
of digital filters used in modern
systems with 10 times faster
sampling rate, Δt_{new}, and
therefore a 10 times shorter
Nyquist period, $T_{NY, new}$. Note
that the filter response, w, is a
complex number describing
both the attenuation and the
phase lag of the output signal
compared to the input.

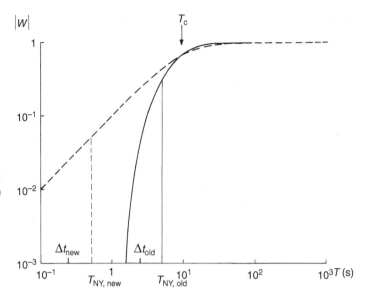

Figure 3.6 Filter response, w, versus period for two long-period data-acquisition systems both having $T_c = 10$ s. Solid curve represents the response of the steep, analogue, anti-alias filters typically used in an old system with sampling rate, Δt_{old}, and Nyquist period, $T_{NY, old}$. Dashed line represents the response characteristics typical of digital filters used in modern systems with 10 times faster sampling rate, Δt_{new}, and therefore a 10 times shorter Nyquist period, $T_{NY, new}$. Note that the filter response, w, is a complex number describing both the attenuation and the phase lag of the output signal compared to the input.

change in capacitance can result in a significant phase shift! On the other hand, now that digital storage media have become relatively inexpensive, we could choose a sampling frequency that is 10 times faster than $\Delta t = T_0/4$ and increase the capacity of the data storage medium by a factor of 10, and then have only one RC circuit with the same cut-off period. In this case, the filter is less steep, and calibration is less critical (Figure 3.6).

Neither of the anti-alias filters shown in Figure 3.6 appears to perform particularly well, since the relative attenuation at the Nyquist period is only 0.3 and 0.05, respectively. However, the intensity of magnetic fluctuations (Figure 1.1) decreases with decreasing period, so that the residual alias will have a small amplitude relative to the signals that are to be analysed. Further details about calibration are given in Chapter 4. Incorrect calibration of data-acquisition systems can lead to severe errors in MT transfer functions, and we sound a cautionary note: never use a data-acquisition system for which an adequate description of the analogue electronics is not available.

Finally, we need to consider what type of storage medium is used for the digital data, and what capacity it should have. Since we have already determined the sampling rate, the storage capacity that is required will depend on the recording duration that is proposed. In general, we should aim to acquire at least 100 samples at the longest period of interest. Therefore, if we are interested in recording the *magnetic daily variation*, we should deploy our instruments for ~100 days. At present, four to five maintenance visits (to check batteries and cables and secure data) would probably be scheduled for a deployment of this duration. Ideally, we might therefore require a data storage medium that could save data from $T_m = 20$ days, and five channels (three components of the magnetic field, and two components of the electric field).

If $T_0 = 8\,s$, $\Delta t = 2\,s$ (assuming a datalogger with 16-bit A/D converter and steep analogue filters) and $T_{\mathrm{m}} = 20$ days (=$1\,72\,8000$ s), a five-channel *time series* will consist of $5 \times T_{\mathrm{m}}/\Delta t = 4\,320\,000$ data points, which require 8.64 Mbytes of memory. If the sample rate is increased by a factor of 10, in order to allow a reduction in the precision with which anti-alias filters are calibrated, then a data storage medium with a capacity of the order 100 Mbytes will be required. Such capacity is readily available on chip cards, which tend to be more reliable under field conditions than mechanically sensitive hard disks. If we design, in contrast, a logger for a broad-band induction coil magnetometer with a period range 0.001–3600 s, then the same line of argument leads us to acquire 1 Gbyte or more of data. However, AMT systems tend to split data recording into *period bands*, and real time AMT systems perform data processing of high-frequency signals simultaneously with data acquisition, reducing the amount of raw data that needs to be stored. It is advisable, however, to store the raw time series for periods longer than ~0.1 s, to enable more rigorous post-processing of data; particularly data in the 1–10 s *dead-band*. In this case, assuming (as before) that we record continuously throughout a site occupation duration that is 100 times longer than the longest period (1 hour), the required data storage capacity will again be 10–100 Mbytes.

3.2 Target area and spatial aliasing

The ideal design for a survey is target specific. In general, shallower targets will demand denser site spacings than deeper targets. However, shallow heterogeneities may distort the way in which deeper structures are imaged (see Chapter 5). As a general rule, it is unwise to interpret anomalies that are only supported by data from one site.

Spatial aliasing is the term used to describe undersampling in the space domain. For remote-sensing techniques this applies to situations in which sites are located too far apart to achieve adequate resolution of the target. *Static shift* (Sections 5.1 and 5.8), is a common manifestation of spatial aliasing.

Figure 3.7 A simple example of spatial aliasing. A sinusoidal signal is sampled at discrete points placed at regular intervals (Δd). The sampling is too sparse for the original signal (solid line) to be retrieved from the sampled data, and the reconstructed signal (dashed line) has a longer wavelength (λ_a) than the wavelength (λ_o) of the original signal. Notice the similarity with aliasing in the time domain (Figure 3.4).

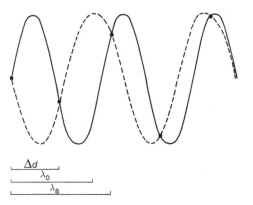

Figure 3.7 A simple example of spatial aliasing. A sinusoidal signal is sampled at discrete points placed at regular intervals (Δd). The sampling is too sparse for the original signal (solid line) to be retrieved from the sampled data, and the reconstructed signal (dashed line) has a longer wavelength (λ_a) than the wavelength (λ_o) of the original signal. Notice the similarity with aliasing in the time domain (Figure 3.4).

A simple example of spatial aliasing is shown in Figure 3.7, in which a sinusoidal signal is sampled at discrete points placed at regular intervals. Because the sampling is too sparse, the original signal cannot be retrieved from the sampled data. The reconstructed signal has a longer wavelength than the original signal. What is the minimum number of samples required to represent the signal shown in Figure 3.7 without loss of information? Compare your answer to the discussion of Nyquist periods in Section 3.1.4.

3.3 Arrays versus profiles: further considerations of dimensionality

Whereas seismologists routinely deploy seismometers in arrays, most MT surveys are still focussed on collecting data along profiles. This is primarily because in the past MT has been adopted as a low-budget remote-sensing technique. However, a number of MT array studies have now yielded results (e.g., Simpson, 2001b; Bahr and Simpson, 2002; Leibecker *et al.*, 2002) that could not have been obtained by deploying instruments along a single profile. Array data have proven particularly useful for constraining electrical anisotropy (see Chapter 9).

Another advantage of array MT data is that it can be used to calculate geomagnetic depth sounding (GDS) transfer functions (Section 10.1) at no extra cost. Joint interpretations of MT and GDS transfer functions provide better constraints than interpretations based on MT transfer functions alone.

With the advent of fast 3-D forward-modelling algorithms (e.g., Mackie *et al.*, 1993), 3-D re-interpretation of MT data from a number of regions where successions of profiles have been surveyed and modelled individually (in 2-D) is a possibility.

3.4 Resolving power and the duration of a campaign

The duration of a field campaign depends, to a large extent, on the availability of instruments and funds. Very often we will have a trade-off between 'many sites' versus 'long occupation of each site'. Field campaigns operated for commercial purposes have generally emphasised maximal site coverage, whereas scientific campaigns for which the objectives involve detailed analysis (Chapter 5, onwards) of high-quality data from individual sites have favoured longer site occupancies. Recently, however, an array of 36 sites was deployed for a sufficiently long duration to perform a conductivity study of the mantle (Leibecker *et al.*, 2002).

The relative *confidence interval* of an impedance, like the standard deviation of a sample, is a linear function of $\sqrt{1/\nu}$, where ν is the number of *degrees of freedom* (i.e., number of independent observations) used in the estimation of the impedance. In Appendix 4 and Section 4.2, we explain that an order of magnitude for the relative error of an impedance, Z, is:

$$\frac{\Delta Z}{Z} = \sqrt{\frac{4 g_{F_{4,\,\nu-4}}(\beta)\varepsilon^2}{\nu - 4}}, \tag{3.6}$$

where $g_{F_{4,\,\nu-4}}(\beta)$ is a statistical measure that depends on ν and on the confidence interval, β (which is a statistical measure of the probability that the predicted value of Z lies within the calculated error). For $\nu > 50$ and $\beta = 68\%$, $g \approx 1.5$. The residuum, ε^2, is the relative fraction of the electric power spectrum that is not correlated with the magnetic field. Experience suggests that, for the longest periods, $\varepsilon^2 \approx 0.15$. Therefore we need $\nu \geq 100$ in order to get $\Delta Z/Z \leq 0.1$. As a conservative estimate, the site occupation time should therefore be at least 100 times the longest sounding period that is of interest. Given that electromagnetic spectra are complex numbers, this will, in theory, provide 200 independent observations, but practically we might discard 50% of the data owing to low signal-to-noise ratios. Of course, more independent observations will be available at shorter periods than at longer periods. For example, if we achieve $\nu = 100$ and $\Delta Z/Z = 0.10$ at the period 24 h, we can achieve $\nu = 2400$ and $\Delta Z/Z = 0.02$ at a period of 1 h. In Chapter 9, we shall discuss a few case studies in which such high-quality data was attained. In other words, if the longest period of interest is the daily variation, we require a 100-day time series. (Don't panic! The instruments can record unattended, and only short visits to read out data every 20 days or so are required).

If a data-acquisition system with induction coils is used to obtain periods spanning 0.001–3600s, then the occupation time should ideally be 4 days. However, modern, real-time, AMT systems are often considered to be too expensive to deploy at one site for so long. Unfortunately, 'economising' on site occupation duration generally results in poor-quality long-period data. A good compromise can be reached by combining the use of an AMT system with a cluster of long-period data-loggers.

3.5 Sources of noise external to the Earth and a preview of processing schemes

Any signals in measured electromagnetic fields corresponding to non-inductive or locally inductive (i.e., of short inductive scale length compared to the skin depth under consideration) sources can be considered to be noise. Static shift (Section 5.1 and Section 5.8) and current channelling (Section 5.9) can each be considered as a manifestation of noise that arises owing to the complexity and inhomogeneity of the real Earth. Such effects induce anomalous currents and charge concentrations, inflicting different and sometimes difficult to quantify distortion effects, and masking the signature of deeper geoelectric structures. Additional sources of noise that lie external to the Earth are of three types:

  (i) cultural;
 (ii) meteorological; and
(iii) sensor.

In populated areas, electricity power lines produce dominant 50 Hz and 150 Hz electromagnetic fields. Whilst noise at such frequencies is relatively easily eliminated by notch filtering, it can limit the dynamic range of magnetic induction coils (see section 3.1.4) and cause instrumental saturation. Power-line noise is highly polarised. Therefore, the effects of power- line noise are usually more prevalent in one orthogonal measurement direction than in the other. Electricity generators can also produce significant levels of noise in the 50 Hz range. Generator noise is harder to eliminate than power-line noise using notch filters owing to its more variable bandwidth. Electric field measurements are also susceptible to contamination from ground leakage currents arising from electric railways and electric fences, the noise spectra from which span broad frequency ranges making filtering difficult.

Automobiles represent a dual source of noise, creating both magnetic and seismic disturbances. Generally, magnetic disturbances can be negated by ensuring that sensors are placed more than 20 m away from any road. Seismic noise, although considerably reduced when the road is founded on firm bedrock, generally exhibits a longer range than magnetic noise. Seismic vibration generates noise on the telluric components by modulating the potential between the electrodes and the ground, and rotational movement, $d\theta$, of the magnetic sensors in the Earth's magnetic field transforms seismic noise into a perturbation of the magnetic field, $\mathbf{B}$ according to:

$$dB = \mathbf{B}[\cos(\theta + d\theta) - \cos\theta], \tag{3.7}$$

where θ is the initial orientation of the magnetic sensor with respect to the Earth's magnetic field. For the worst case scenario, which occurs when $\theta = 90°$, a rotation of only $0.002°$ generates a perturbation of 1 nT.

A ubiquitous source of meteorological noise is wind. The vibration of telluric lines in the wind can generate voltages comparable to short-period telluric signals. As a result of wind blowing on trees and bushes their roots may move within the Earth, generating seismic noise, which may, in turn, cause movement of the sensors and corresponding perturbations of the measured fields. High-frequency vertical magnetic field measurements are generally worst effected by wind vibration and ground roll. Another source of meteorological noise is generated by local lightning discharges, which superimpose noise on the source field owing to their inhomogeneous and impulsive nature. Lightning may also cause saturation of telluric amplifiers.

Sensor noise and noise arising from electronic circuitry is usually independent of signal power and random in nature, making it difficult to distinguish and harder to evaluate. However, sensor noise is generally low (e.g., less than 30 pT for a fluxgate magnetometer) for modern instrumentation. As discussed in Section 3.1.3, care should be exercised to ensure that the effects of temperature variations on the sensors, and electronic components are minimised. This can be achieved by burying sensors as deeply as possible, and choosing a shady place for the datalogger.

At periods exceeding 1000 s, the signal-to-noise ratio is independent of signal power (Egbert and Booker, 1986), but at shorter periods, for which the power of the natural source field is more variable and contamination by cultural noise is more prevalent, signal-to-noise ratios can vary significantly. The presence of noise causes bias effects, including false depression or enhancement of calculated impedance tensors (see Chapter 4).

The level of random noise that is present in data can be quantified by considering the amount of linear correlation between electric and magnetic field components. The correlation coefficient is called *coherence* ψ, and is expressed as a spectral ratio composed of the cross-correlated electric and magnetic field spectra $\langle E^*B \rangle$ (where * denotes complex conjugate, and E and B are individual electric and electromagnetic field componenents) that are used to calculate the transfer function (see Chapter 4), divided by their two auto-power spectra:

$$\psi = \frac{\langle E^*B \rangle}{(\langle E^*E \rangle \langle B^*B \rangle)^{1/2}} \tag{3.8}$$

Coherence is a dimensionless real variable with values in the range $0 \leq \psi \leq 1$. The upper limit of 1 is indicative of perfectly coherent signals. With modern equipment and data processing schemes (Chapter 4), MT transfer functions with $\psi > 0.9$ can be produced routinely.

Coherence threshold values are often applied as pre-selection criteria when recording AMT data. Relying purely on coherence values to identify noise can, however, fail in multi-dimensional environments, owing to noise being correlated across different electromagnetic components.

Figure 3.8 shows transfer functions in the 32–170 s period range calculated from a single time series window of length 1024×2 s samples (i.e., 2048 s). Assuming an ideal level of activity during the chosen data window, the maximum number of 32-s independent observations that can be recorded is 64, whilst the maximum number of 170-s independent observations is 12. The actual number of independent observations centred on each period is shown in Figure 3.8(c). The coherence (Equation (3.8)) is approximately the same at 32 s and 170 s (Figure 3.8(d)). Therefore, the errors of the impedance increase by a factor that depends on the square root of the factor by which the number of independent observations is reduced (see Equation (3.6)). At 32 s, there are 44 recorded independent observations, whereas at 170 s there are only 10 recorded independent observations. Therefore, the number of independent observations at 32 s is approximately 4 times the number of independent observations at 170 s, and the confidence intervals of the apparent resistivity (which is of similar magnitude at both periods) are approximately twice as large at 170 s as those at 32 s. The dependence of the errors of the transfer functions on both coherence and number of independent observations generally results in a trade-off during processing, between accepting an adequate number of events versus

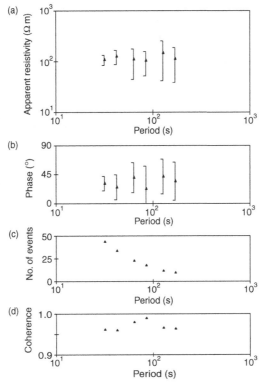

Figure 3.8 (a) Apparent resistivity and (b) impedance phase transfer functions and their 68% confidence intervals at periods spanning 32–170 s calculated from a single time series window containing 1024 samples with a sampling rate of 2 s. The number of independent observations (events) for each period is shown in (c), and their coherence is shown in (d). The spread of confidence intervals depends on both the coherence and number of independent observations. Confidence intervals of the impedance phases (b) scale differently from those of the apparent resistivities (a) (compare Equations (A4.21) and (A4.22)).

selecting only those events with high coherence. For example, the highest coherence of 0.99 occurs at ∼85 s. In spite of this peak in coherence, the error bars are larger at 85 s than at 32 s, because there are only 18 independent observations at 85 s compared to 44 independent observations at 32 s. On the other hand, there are 23 independent observations at 64 s (neighbouring period to 85 s) compared to only 18 independent observations at 85 s, but the confidence intervals of the apparent resistivity at 85 s are smaller than those at 64 s because of the higher coherence (0.99 at 85 s compared to 0.98 at 64 s). Clearly, the errors could be reduced across the entire period span by *stacking* events from more than one time series window. The trade-off between coherence and the number of independent observations is described mathematically (Equation (4.19)) in Section 4.2.

3.6 Economic considerations

In practice, financial considerations may make it unfeasible to conduct an ideal field campaign in one field season. If funding is limited, should you occupy many sites for shorter time intervals, or fewer sites for more time?

The ideal occupation time at a measurement site will depend on the target and hence the frequency range of measurements under consideration. The conductance of an anomaly can only be constrained if completely penetrated. In general, it will be better to acquire good-quality data at fewer sites than poor-quality data at many sites. Good-quality data can be expanded upon later by increasing site density or expanding the region surveyed, as necessary, following an initial interpretation. On the other hand, over-stretching resources by cutting site occupation time in order to acquire data from as many sites as possible is likely to result in poor data and mean that sites have to be re-occupied subsequently. Time should always be taken to assess data quality in the field before relocating to another site.

3.7 Suggested checklist of field items

Each MT station will typically consist of the following:

One datalogger
One fluxgate magnetometer or three magnetic induction coils
Four electrodes
Four telluric cables
One magnetometer cable or three magnetic induction coil cables
One metal stake to earth
One Global Positioning System (GPS) receiver (optional)
One power source (e.g., batteries)

A notebook computer can be used for downloading data from the datalogger and making in-field data processing. In addition, the following accessories will be of use during installation and servicing of the MT site: compass, measuring tape, spirit level, spare cables, digging tools, siting poles, spare data storage media, spare batteries, battery charging devices, insulating tape, digital multimeter, pencil and paper, zip disks (or similar) for archiving data.

3.8 A step-by-step guide to installing an MT station

A typical site layout is shown in Figure 3.9. Trees and bushes provide shelter from the Sun and prying eyes, and can therefore be good places to place the box containing the datalogger and batteries.

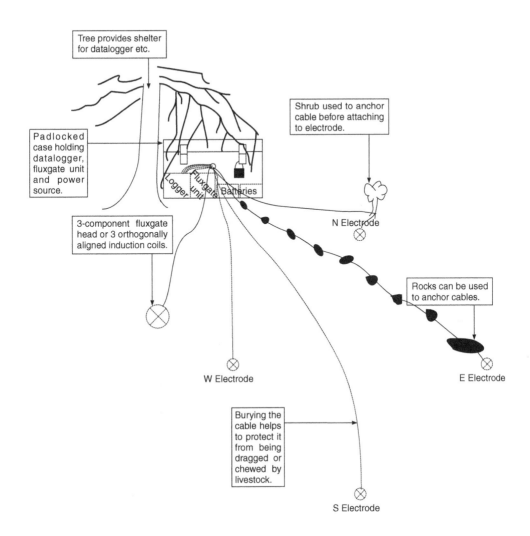

Tree provides shelter for datalogger etc.

Padlocked case holding datalogger, fluxgate unit and power source.

Shrub used to anchor cable before attaching to electrode.

Logger   Fluxgate unit   Batteries

3-component fluxgate head or 3 orthogonally aligned induction coils.

N Electrode

Rocks can be used to anchor cables.

W Electrode

E Electrode

Burying the cable helps to protect it from being dragged or chewed by livestock.

S Electrode

Figure 3.9 Typical layout of an MT station.

Depending on digging conditions and intended recording time, it may be worthwhile to shield the datalogger further from temperature variations by burial or coverage with fallen branches, etc.

Having chosen a suitable position for the datalogger, the positions of the electrodes can be sited. This can be achieved accurately using siting poles and a geological compass. Some MT practitioners advocate measuring a standard dipole length (e.g., 50 m) during the siting process, whereas others estimate the distance between electrodes and measure the actual distances subsequently. It can be difficult to obtain accurate measurements of the electrode separations using a conventional measuring tape, particularly if there is vegetation to be circumvented. However, in practice, a high degree of accuracy is not necessary, since minor errors contribute an insignificant static-type shift to the impedance magnitudes. Dipoles are normally aligned in the north–south (N–S) and east–west (E–W) directions, and may be configured to form a + or an L. If a + is chosen, then four electrodes are required, whereas for an L only three electrodes are required, because one electrode is common to both dipoles. If the Earth were 1-D, then the + configuration would have the advantage that if one of the dipoles were to fail (owing, for example, to a cow chewing through the cable), then electric fields would still be registered for the other dipole, and 1-D modelling would still be viable. An advantage of the L-configuration is that having a common electrode makes parallel connection (Figure 3.10) of telluric lines impossible. The risk of connecting telluric lines to the

Figure 3.10 (a) Field configuration in which electrodes (⊗) are connected to a datalogger (box) to establish a pair of orthogonal dipoles that form a +. (b) Accidental jumbling of the N and E telluric cables results in two parallel dipoles oriented NE–SW. (c) Field configuration in which electrodes are connected to a datalogger to establish a pair of orthogonal dipoles that form an L. The S and W telluric inputs are inter- connected at the datalogger. (d) Accidental jumbling of the telluric cables results in two dipoles that are oriented at 45° to each other. In this case, orthogonal directions of the electric field can be retrieved mathematically.

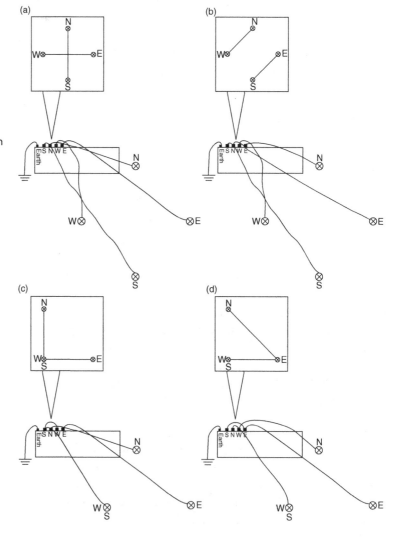

incorrect input is also minimised by labelling telluric cables according to their directions as they are laid out.

The electrodes should be buried below the surface to a depth that mitigates temperature variations (see Section 3.1.3). Waterproof plugs should be used to connect electrode wires to the telluric cables. We recommend anchoring the telluric cables, particularly close to where they are connected to the electrodes. This can be achieved by tying cables to a shrub or rock, etc. (Figure 3.9). It can be a good idea to measure the contact resistance of the electrodes using a voltmeter. If contact resistances exceed the $k\Omega$ range, then this may indicate a poor contact (or a dried-out electrode). A metal stake can be used to earth the telluric inputs. After connecting the telluric cables to the appropriate inputs of the datalogger, the voltages between paired electrodes can be compensated. The appropriate level of the pre-amplifier should also be selected.

The fluxgate magnetometer, or induction coils should be buried at least 5 m from the datalogger. If induction coils are used, then it is necessary to orientate them using a compass, and to level them using a spirit level. If a fluxgate magnetometer is used then the fluxgate head should be approximately orientated and levelled. Thereafter, fine adjustments can be made using screw mechanisms (Figure 3.1). The fluxgate head is connected via a coaxial cable to an electronic unit from which it receives power, and to which it transmits its output signal. Compensation (see section 3.1.4) may also be facilitated via this unit, which is achieved by centring three needles (one for each magnetic component). Typically, fine adjustments to the fluxgate head are made until the y-component of the magnetic field requires zero compensation, and a spirit level indicates that the head is level. Subsequently, the x- and z-components are compensated at the electronic unit.

With the sensors in place, information such as site name, pre-amplifier values and telluric lengths can be entered into the datalogger. This information will then be available in the header of the datafile to be produced. The date and universal time (UT) should also be correctly set (e.g., using a Global Positioning System (GPS) receiver).

You are ready to record data.

Chapter 4

From time series to transfer functions: data processing

The digital time series collected during an MT survey can easily total a few Gigabytes, but the data that will finally be interpreted with numerical modelling schemes typically consist of a few hundred numbers per site that represent the frequency-dependent transfer functions. The reduction is referred to as 'data processing'. One time series can simultaneously contain information about many periods and, therefore, about many penetration depths, and the first step in data processing involves a Fourier transformation from the time domain to the frequency domain. Data reduction is then achieved by stacking data falling within particular spectral bands in the frequency domain: both neighboured frequencies from the same segment (window) of a time series, and similar frequencies from sequential time series windows can be stacked.

What is the exact meaning of 'transfer function'? The Earth is regarded as a linear system that responds to an input process (e.g., a time-varying magnetic field) via a predictable output process, (e.g., a time-varying electric field). The transfer function is the ratio of these processes, and because the system is linear, the transfer function does not depend on the amplitude of the input process. The estimation of the transfer functions can be hindered by noise in the input and output processes: e.g., electric and magnetic fields which are not naturally induced but are rather associated with our technical civilisation. If the noise level is low, or the noise has a Gaussian distribution (the frequency of occurrence, plotted as function of the size of the noise, delineates a Gaussian bell), then a least-square estimation (in which the squared residuum, which is not explained by the linear system, is minimised) is sufficient. Otherwise, the occurrence of outliers might force us to apply a 'robust' processing scheme. Most robust processing schemes operate in an iterative manner, and use some measure of the departure of an individual contribution from the average to down-weight outliers in the next iteration. What is an outlier? As an example, suppose that an attempt is made to estimate the average height of pupils in a school class, but the sample includes four children (1.0, 1.1, 1.1, 1.2 m, respectively) and a basketball player (2.0 m). In this example, the basketball player would be considered to be an outlier. We finally discuss a few ways of displaying the transfer functions, as well as the advantages and disadvantages of various visualisation aids.

4.1 Fourier transformation, calibration and the spectral matrix

Figure 4.1 shows an example of MT *time series* collected at a site in central Australia. The sampling rate was $\Delta t = 2$ s and the *time window* shown in Figure 4.1 is 30 minutes long. Therefore, 900 data points are plotted for each component. As there are five components and the data are 16 bit (2 byte), this 30-minute time window represents 9 kbytes of data. An electromagnetic time series recorded over a few weeks or months, can easily generate 10 Mbytes of data (see Section 3.1.4). In contrast, the *transfer functions* of one site are a very small dataset, typically described by the complex *impedance tensor* at 30–50 evaluation frequencies. The data reduction is achieved in two ways:

Figure 4.1 30-minute segment of five-component MT time series recorded at 'Mt Doreen' in central Australia.

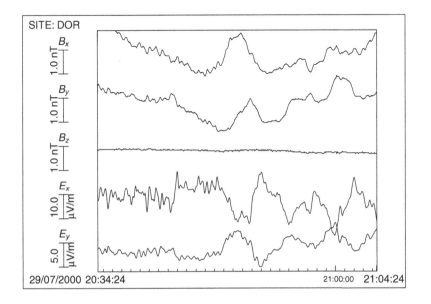

SITE: DOR

B_x 1.0 nT

B_y 1.0 nT

B_z 1.0 nT

E_x 10.0 μV/m

E_y 5.0 μV/m

29/07/2000 20:34:24   21:00:00   21:04:24

(i) *Stacking* of observations from discretised time windows. (More precisely, the power and cross spectra are stacked in the frequency domain, as we shall see later in this section).

(ii) Weidelt's dispersion relations (Equation (2.37)) predict that neighbouring frequencies will provide similar transfer functions, allowing us to perform a *weighted stacking* of information from neighbouring frequencies.

We next describe a simple but flexible processing scheme that can be applied either automatically or with user-selected time windows. It can be operated in conjunction with a *least-square* estimation of the transfer function, or incorporate *robust processing* or *remote reference processing* of noisy data (Section 4.2).

Step 1: calculating the Fourier coefficients of one interval of a single time series

Suppose, for example, that a digital time series that represents the B_x-component has been cut into intervals of N data points, each of which can be denoted as x_j, where $j = 1, N$ (in Figure 4.1, $N = 900$). Then, by applying the *trend removal* and the *cosine bell*, and the *discrete Fourier transformation,* (Equations (A5.12), (A5.13) and (A5.10), explained in Appendix 5), we obtain $N/2$ pairs of Fourier coefficients a_m, b_m which are combined as *complex Fourier coefficients*

$$\tilde{X}_m = a_m + ib_m, \tag{4.1}$$

where

$$\left.\begin{array}{l} a_m = \dfrac{2}{N} \sum_{n=0}^{N-1} x_j \cos m\phi_j \\[2ex] b_m = \dfrac{2}{N} \sum_{j=0}^{N-1} x_j \sin m\phi_j \end{array}\right\} m = 1, \ldots, N/2.$$

In the following, we anticipate that all energy above the *Nyquist frequency*, ω_{NY}, has been removed from the analogue time series by an anti-alias filter (as described in Section 3.1.4) prior to digitisation. Therefore,

$$\tilde{x}(\omega) = 0, \qquad \forall\, \omega > \omega_{\mathrm{NY}} = \frac{2\pi}{2\Delta t} \tag{4.2}$$

and the discrete Fourier transformation (Equation (A5.10)) recovers the entire information contained in the digital time series x_j, $j = 1, N$.

In the following discussion, we will denote the mth discrete *spectral lines* in the *raw spectra* of B_x, B_y and B_z, and E_x and E_y as $\tilde{X}_m, \tilde{Y}_m$ and $\tilde{Z}_m$, and $\tilde{N}_m$ and $\tilde{E}_m$, respectively.

Step 2: calibration

Because a computer can only store integer numbers, a relationship must be established between the output of a sensor and the number stored in a data-acquisition system. This step is called *calibration*[5], and consists of four processes.

1 Frequency-independent calibration of the sensors.

   We need to know the *sensitivity* of the sensors. For magnetic sensors, the sensitivity has the unit $V\,T^{-1}$ or, more usually, $mV\,nT^{-1}$. For example, the *fluxgate magnetometer* in Figure 3.2 has a sensitivity $S_B = 5\ mV\ nT^{-1}$. For electric amplifiers, the sensitivity has the unit $mV\ (mV/m)^{-1}$ and is calculated from the dimensionless amplification factor A of the amplifier and the distance, d, between the electrodes:

$$S_E = A \times d \qquad (4.3)$$

2 Frequency-dependent *calibration* of the sensors.

   Some sensors, such as the *induction coil* in Figure 3.2 have a frequency-dependent sensitivity. In this case, step 1 should be performed at individual frequencies only. The frequency-dependent sensitivity is described by a dimensionless complex factor $w_m = w(f_m)$, where f_m is the mth frequency in the output of the Fourier transformation. Any anti-alias filter also has a frequency-dependent complex calibration coefficient, (which is ideally 0 at the Nyquist frequency and 1 at the first evaluation frequency). There are two ways of avoiding the necessity of calibrating anti-alias filters.

   (i) In the modern design of instrument described in Figure 3.6 the Nyquist frequency and the first evaluation frequency are so far away from each other that $w = 1$ at the first evaluation frequency.

   (ii) The hardware design of the anti-alias filters of all components is the same and so accurate that estimations of transfer functions (which are ratios of different components of the electromagnetic field) are not affected.

---

[5] The term *calibration* is also used to denote estimation of calibration factors related to filters.

3 Calibration of the A/D converter.

We need to know the resolution of the A/D converter: e.g., a 16-bit A/D converter that can take any voltage in the $-5V$, $+5V$ range has a *least count* of l.c. $= 10\,\text{V}/2^{16} = 0.153\,\text{mV/bit}$.

4 Application of the calibration coefficient to the output of the Fourier transformation.

In principle, the calibration factors from processes 1 and 3 can be applied to the digital data at any stage in the processing, but obviously the frequency-dependent calibration factor w_m from process 2 can only be applied following the Fourier transformation. Some electronic equipment (e.g., a steep anti-alias filter) is described by convolution of data with a function $w(f)$, and this convolution should be performed on the raw spectrum, before the frequency resolution is reduced in step 3:

$$\tilde{X}_m \rightarrow l.c. \times w(f_m)^{-1} \times S_B^{-1} \times \tilde{X}_m, \quad \text{and}$$
$$\tilde{N}_m \rightarrow l.c. \times w(f_m)^{-1} \times S_E^{-1} \times \tilde{N}_m. \tag{4.4}$$

Step 3: The evaluation frequencies and the spectral matrix

The choice of evaluation frequencies (or periods) is somehow arbitrary, but two conditions apply:

1 The evaluation frequencies (or periods) should be equally spaced on a logarithmic scale. For example, if we choose 10 s and 15 s as evaluation periods, then we should also choose 100 s and 150 s (rather than 100 s and 105 s), because the relative error dp/p of the *penetration depth* (Equation (2.20)) is then the same for periods of the order 10 s and 100 s.

2 Ideally, we should have 6–10 evaluation frequencies per decade – more are unnecessary because Weidelt's dispersion relation (Equation (2.37)) predicts similar results for neighbouring frequencies, but fewer could result in an *aliasing* in the frequency domain.

A suggested frequency distribution is:

$$f_1 = f_{\max}$$
$$f_2 = f_{\max}/\sqrt{2}$$
$$f_3 = f_{\max}/2$$
$$\vdots$$
$$f_k = f_{\max}/2^{\sqrt{(k-1)}}. \tag{4.5}$$

These evaluation frequencies were used to produce Figure 4.2, with $f_{\max} = 0.125\,\text{Hz}$ or $T_{\min} = 8\,\text{s}$.

In Figure 4.1, the sample rate is 2 s and the shortest period is 4 s and therefore, $f_{\max} = 0.25\,\text{Hz}$. Practically, however, we want at

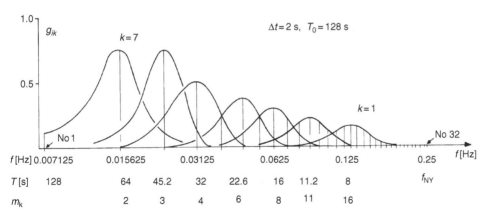

Figure 4.2 Relationship between the raw spectrum (output of the Fourier transformation) and the evaluation frequencies used for computing MT transfer functions. The form of the spectral window is chosen to optimise the conflicting demands of number of degrees of freedom (which favours including as much data as possible) versus data resolution (which favours not smoothing over too many discrete frequencies). The spectral window weight, g_{ik} is defined by Equation (4.7). The m_k frequencies occurring in the raw spectrum are equi-spaced on a linear scale, whereas the k evaluation frequencies are equi-spaced on a logarithmic scale. Therefore, more smoothing occurs at short periods.

least four samples per period and therefore $f_{max} = 0.125\,$Hz. The lowest frequency (longest period) is determined by the window length. Typically, we might use a window consisting of 512 data points, but for ease of illustration, the data reduction principle is demonstrated in Figure 4.2 assuming a 128 s time window (i.e, 64 discrete data points), for which $f_k = 1/64\,$Hz and $f_{max}/f_k = 8$. The raw spectrum from which Figure 4.2 is produced includes $M = 32$ frequencies, whereas $k \approx 7$ evaluation frequencies result from Equation (4.5). (The first line in the raw spectrum should be ignored, as it is affected by the *cosine bell* (Equation (A5.13)). Averaging is performed in the frequency domain. At higher frequencies, more information is merged into the same evaluation frequency in order to meet the requirement that the evaluation frequencies are equally spaced on a logarithmic scale. (It could also be stated that, at short periods, we have more independent information per period of the resulting transfer function). The degree of averaging and the form of the *spectral window* shown in Figure 4.2 is a compromise between the principle of incorporating as much data as possible (because then the number of degrees of freedom, which determines the confidence intervals of the transfer functions, will be larger) and not smoothing too much (because mixing different evaluation periods that are associated with different penetration depths reduces resolution of individual conductivity structures). In other words, the aim is to reach a trade-off between data errors and resolution. In Figure 4.2, a Parzen window (Parzen, 1961; Jenkins and Watts, 1968; Parzen, 1992) is used, but this is only a suggestion.

We next provide an overview of a possible procedure for estimating the power and cross spectra (Equations (A6.3) and (A6.4)) in

Appendix 6). For the kth evaluation frequency, we need to know the number of raw data to be merged:

$$j_k = f_k/2\Delta f = \frac{T \cdot f_k}{2},$$ (4.6)

where Δf is the space between the lines of the raw spectrum and $T = 1/\Delta f$ is the span of the *time window*. For the Parzen window, the ith spectral window weight at the kth evaluation frequency is:

$$g_{ik} = g_{0k}\left[\frac{\sin(i\pi/j_k)}{i\pi/j_k}\right]^4,$$ (4.7)

where the scaling constant $g_{0k} = 1/j_k$ has been chosen in such a way that:

$$g_{0k} + 2\sum_{i=1}^{j_k} g_{ik} = 1.$$ (4.8)

The number of the central frequency of the window is:

$$m_k = f_k/\Delta f.$$ (4.9)

The numerical integration depicted in Figure 4.2 is therefore,

$$\langle \tilde{X}\tilde{X}^* \rangle_k = g_{0k}\tilde{X}_{m_k}\tilde{X}^*_{m_k} + \sum_{i=1}^{j_k} g_{ik}\tilde{X}_{m_k-i}\tilde{X}^*_{m_k-i} + \sum_{i=1}^{j_k} g_{ik}\tilde{X}_{m_k+i}\tilde{X}^*_{m_k+i},$$ (4.10)

where $\tilde{X}^*$ denotes the complex conjugate of $\tilde{X}$. Equation (4.10) is a numerical estimation of the power spectrum of the B_x component at the kth evaluation frequency. Similar equations can be used to compute other power spectra and cross spectra from the data. These spectra are then stored in a *spectral matrix* for each evaluation frequency k:

$$
\begin{array}{ccccc|c}
\tilde{X} & \tilde{Y} & \tilde{Z} & \tilde{N} & \tilde{E} & \\
\hline
\tilde{X}\tilde{X}^* & \tilde{Y}\tilde{X}^* & \tilde{Z}\tilde{X}^* & \tilde{N}\tilde{X}^* & \tilde{E}\tilde{X}^* & \tilde{X} \\
 & \tilde{Y}\tilde{Y}^* & \tilde{Z}\tilde{Y}^* & \tilde{N}\tilde{Y}^* & \tilde{E}\tilde{Y}^* & \tilde{Y} \\
 & & \tilde{Z}\tilde{Z}^* & & & \tilde{Z} \\
 & & & \tilde{N}\tilde{N}^* & \tilde{E}\tilde{N}^* & \tilde{N} \\
 & & & & \tilde{E}\tilde{E}^* & \tilde{E}.
\end{array}
$$ (4.11)

The number of independent observations or degrees of freedom is

$$\nu_k = 2(1+2)\sum_{i=1}^{j_k} g_{ik}/g_{0k}.$$ (4.12)

If more than one time window is evaluated (which, as mentioned in connection with stacking at the beginning of this section, is normally the case), then the power and cross spectra as well as the number of degrees of freedom from a sequence of time windows are stacked to the previously evaluated data, e.g.,:

$$\langle\langle \tilde{X}\tilde{X}^* \rangle\rangle = \sum_{l=1}^{\text{all windows}} \langle \tilde{X}\tilde{X}^* \rangle_l. \tag{4.13}$$

A proper estimation of a power spectrum (see Appendix 6 for mathematical details thereof) would require some sort of normalisation, but for the estimation of transfer functions we are only interested in spectral **ratios**. The stacking procedure can be modified by attributing a weight to the spectra from each particular window. The weights can be found by hand-editing (in this case they are usually 0 or 1) or with a robust technique, as described in the following section. Recently, Manoj and Nagarajan (2003) applied artificial neural networks in order to find the stacking weights.

4.2 Least square, remote reference and robust estimation of transfer functions

Before the full potential of MT data can be realised it is generally necessary to remove biasing effects arising from noise (Section 3.5). This stage is referred to as data processing. During data processing, *earth response functions* or transfer functions (such as *impedance tensors*), which are more intuitively meaningful and exhibit less source field dependency than *time series* are also calculated.

Linear expansion of Equation (2.50) yields the pair of equations:

$$E_x(\omega) = Z_{xx}(\omega)H_x(\omega) + Z_{xy}(\omega)H_y(\omega) \tag{4.14a}$$

$$E_y(\omega) = Z_{yx}(\omega)H_x(\omega) + Z_{yy}(\omega)H_y(\omega). \tag{4.14b}$$

Equations (4.14a) and (4.14b) are made inexact by measurement errors, and because the *plane wave* source field assumption is only an approximation. Therefore, a statistical solution for the MT impedance, $\underline{Z}$, is required that minimises a remainder function, $\delta Z(\omega)$. The remainder function represents uncorrelated noise. For the case that $\delta Z(\omega)$ represents electrical noise, it is added to the right-hand sides of Equations (4.14a) and (4.14b):

$$E_x(\omega) = Z_{xx}(\omega)H_x(\omega) + Z_{xy}(\omega)H_y(\omega) + \delta Z(\omega) \tag{4.15a}$$

$$E_y(\omega) = Z_{yx}(\omega)H_x(\omega) + Z_{yy}(\omega)H_y(\omega) + \delta Z(\omega). \tag{4.15b}$$

Least-square and robust processing techniques are examples of statistical processing methods commonly employed to remove noise from MT data. In both techniques, Equation (4.15) is solved as a *bivariate linear regression* problem (see Appendix 4). Least-square processing techniques involve isolating the components of $\underline{Z}$ using least-square cross-power spectral density estimates, with the function $\delta Z(\omega)$ assumed to have a *Gaussian (normal) distribution*. Cross-powers for a discrete frequency can be generated in the frequency domain by multiplying Equations (4.14a) and (4.14b) by the complex conjugates (denoted *) of the electric and magnetic spectra as follows:

$$\langle E_x(\omega)E_x^*(\omega)\rangle = Z_{xx}(\omega)\langle H_x(\omega)E_x^*(\omega)\rangle + Z_{xy}(\omega)\langle H_y(\omega)E_x^*(\omega)\rangle \quad (4.16a)$$

$$\langle E_x(\omega)E_y^*(\omega)\rangle = Z_{xx}(\omega)\langle H_x(\omega)E_y^*(\omega)\rangle + Z_{xy}(\omega)\langle H_y(\omega)E_y^*(\omega)\rangle \quad (4.16b)$$

$$\langle E_x(\omega)H_x^*(\omega)\rangle = Z_{xx}(\omega)\langle H_x(\omega)H_x^*(\omega)\rangle + Z_{xy}(\omega)\langle H_y(\omega)H_x^*(\omega)\rangle \quad (4.16c)$$

$$\langle E_x(\omega)H_y^*(\omega)\rangle = Z_{xx}(\omega)\langle H_x(\omega)H_y^*(\omega)\rangle + Z_{xy}(\omega)\langle H_y(\omega)H_y^*(\omega)\rangle \quad (4.16d)$$

$$\langle E_y(\omega)E_x^*(\omega)\rangle = Z_{yx}(\omega)\langle H_x(\omega)E_x^*(\omega)\rangle + Z_{yy}(\omega)\langle H_y(\omega)E_x^*(\omega)\rangle \quad (4.16e)$$

$$\langle E_y(\omega)E_y^*(\omega)\rangle = Z_{yx}(\omega)\langle H_x(\omega)E_y^*(\omega)\rangle + Z_{yy}(\omega)\langle H_y(\omega)E_y^*(\omega)\rangle \quad (4.16f)$$

$$\langle E_y(\omega)H_x^*(\omega)\rangle = Z_{yx}(\omega)\langle H_x(\omega)H_x^*(\omega)\rangle + Z_{yy}(\omega)\langle H_y(\omega)H_x^*(\omega)\rangle \quad (4.16g)$$

$$\langle E_y(\omega)H_y^*(\omega)\rangle = Z_{yx}(\omega)\langle H_x(\omega)H_y^*(\omega)\rangle + Z_{yy}(\omega)\langle H_y(\omega)H_y^*(\omega)\rangle. \quad (4.16h)$$

Most of these Equations (e.g., Equation (4.16a)) contain auto-powers. Since any component will be coherent with itself, any noise present in that component will be amplified in the auto-power, causing Z_{ij} to be biased. One solution to this problem is the *remote reference method*.

The remote reference method (Goubau *et al.*, 1979; Gamble *et al.*, 1979; Clarke *et al.*, 1983) involves deploying additional sensors (usually magnetic) at a site removed (remote) from the main (local) measurement site. Whereas the uncontaminated (natural) part of the induced field can be expected to be coherent over spatial scales of many kilometres, noise is generally random and incoherent. Therefore, by measuring selected electromagnetic components at both local and remote sites, bias effects arising from the presence of noise that is uncorrelated between sites can be removed. Correlated noise that is present in both local and remote sites cannot be removed by this method. The distance between local and remote sites that is necessary in order to realise the assumption of uncorrelated noise depends on the noise source, intended frequency range of measurements and conductivity of the sounding medium. In the past, the separation distance between local and remote sites was often determined by equipment limitations, owing to the requirement that local and remote sites should be linked by cables. However, with the advent of GPS clocks, which allow instruments to be precisely synchronised, simultaneous data can be ensured without the need for cable links, and local and remote instruments can be placed many kilometres apart. Magnetic remote reference is favoured over electric, because horizontal magnetic fields exhibit greater homogeneity than electric fields in the vicinity of lateral heterogeneities, are less susceptible to being polarised, and are generally less contaminated by noise than electric fields. Using the notation from Section 4.1 and adopting subscripts r to denote the remote reference magnetic fields, Equations (4.16) can be solved for Z_{ij} giving:

$$Z_{xx} = \frac{\langle \tilde{N}\tilde{X}_r^* \rangle \langle \tilde{Y}\tilde{Y}_r^* \rangle - \langle \tilde{N}\tilde{Y}_r^* \rangle \langle \tilde{Y}\tilde{X}_r^* \rangle}{DET}$$

$$Z_{xy} = \frac{\langle \tilde{N}\tilde{Y}_r^* \rangle \langle \tilde{X}\tilde{X}_r^* \rangle - \langle \tilde{N}\tilde{X}_r^* \rangle \langle \tilde{X}\tilde{Y}_r^* \rangle}{DET}$$

$$Z_{yx} = \frac{\langle \tilde{E}\tilde{X}_r^* \rangle \langle \tilde{Y}\tilde{Y}_r^* \rangle - \langle \tilde{E}\tilde{Y}_r^* \rangle \langle \tilde{Y}\tilde{X}_r^* \rangle}{DET}$$

$$Z_{yy} = \frac{\langle \tilde{E}\tilde{Y}_r^* \rangle \langle \tilde{X}\tilde{X}_r^* \rangle - \langle \tilde{E}\tilde{X}_r^* \rangle \langle \tilde{X}\tilde{Y}_r^* \rangle}{DET}, \tag{4.17}$$

where

$$DET = \langle \tilde{X}\tilde{X}_r^* \rangle \langle \tilde{Y}\tilde{Y}_r^* \rangle - \langle \tilde{X}\tilde{Y}_r^* \rangle \langle \tilde{Y}\tilde{X}_r^* \rangle.$$

Now, in order to calculate $\underline{Z}$ at a particular frequency, we simply substitute the appropriate power spectra and cross spectra, which we computed in Section 4.1 and stored in the spectral matrix described by Equation (4.11), into Equations (4.17). The impedance tensor calculated from Equation (4.17) will not be biased by noise provided that the noise in the local measurements is uncorrelated with the noise recorded by the remote reference configuration, but a high degree of correlation between the naturally-induced electromagnetic fields at the local and remote sites is required. In multi-dimensional environments noise may be exaggerated in the cross-spectral density terms, as well as in the auto-power terms. Recently, Pádua *et al.* (2002) showed that remote reference methods may be no better than single-station processing techniques at removing the kind of noise generated by DC railways.

A more sophisticated use of a remote reference site involves a two-source processing technique, in which the electric field expansion (Equation (4.15)) includes a second magnetic field (which represents the noise in the local magnetic field) and a second impedance (which represents the noisy part of the electric fields that is correlated with the magnetic noise). By using the remote reference site to separate these two sources of noise, it is possible under certain circumstances to remove noise which is correlated between the magnetic and the electric readings at the local site (Larsen *et al.*, 1996). An instructive field example is given by Oettinger *et al.*, (2001).

The elements of the induction arrow, $T(\omega)$ (Equation (2.49)), are estimated similarly to $Z(\omega)$ (i.e., by multiplying by the complex conjugates of the horizontal magnetic fields to give):

$$\left.\begin{array}{l} T_x = \dfrac{\langle \tilde{Z}\tilde{X}^*\rangle\langle \tilde{Y}\tilde{Y}^*\rangle - \langle \tilde{Z}\tilde{Y}^*\rangle\langle \tilde{Y}\tilde{X}^*\rangle}{DET} \\[4mm] T_y = \dfrac{\langle \tilde{Z}\tilde{Y}^*\rangle\langle \tilde{X}\tilde{X}^*\rangle - \langle \tilde{Z}\tilde{X}^*\rangle\langle \tilde{X}\tilde{Y}^*\rangle}{DET} \end{array}\right\}. \tag{4.18}$$

MT transfer functions should vary smoothly with frequency (Weidelt, 1972). Therefore, if large changes in MT transfer functions are apparent from one frequency to the next it may be inferred that they have been inaccurately estimated.

We calculate the confidence intervals of the transfer function as detailed in Appendix 4 (Equation (A4.20)) with a modification that since two **complex** transfer functions have been estimated, the term $2/(\nu - 2)$ is replaced by $4/(\nu - 4)$. For example,

$$\begin{array}{l} |\Delta T_x|^2 \le \dfrac{g_{F_{4,\nu-4}}(\beta)4\varepsilon^2\langle \tilde{Z}\tilde{Z}^*\rangle\langle \tilde{Y}\tilde{Y}^*\rangle}{(\nu - 4)DET} \\[4mm] |\Delta T_y|^2 \le \dfrac{g_{F_{4,\nu-4}}(\beta)4\varepsilon^2\langle \tilde{Z}\tilde{Z}^*\rangle\langle \tilde{X}\tilde{X}^*\rangle}{(\nu - 4)DET}, \end{array} \tag{4.19}$$

where the residuum is estimated according to Equation (A4.17):

$$\varepsilon^2 = 1 - \psi^2 = 1 - \frac{T\langle \tilde{Z}\tilde{X}^*\rangle^* + T\langle \tilde{Z}\tilde{Y}^*\rangle^*}{\langle \tilde{Z}\tilde{Z}^*\rangle}. \tag{4.20}$$

Can you prove that the bivariate coherency ψ^2 is real?

Fundamental to least-square statistical estimation is the assumption of power independent, uncorrelated, identically distributed Gaussian errors, whereas Egbert and Booker (1986) have shown that deviations from a Gaussian error distribution occur owing to error magnitudes being proportional to signal power, and owing to

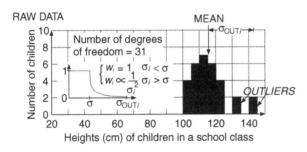

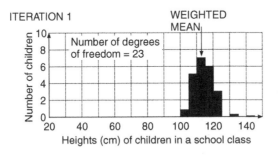

Figure 4.3 Trivial example of how robust processing can be used to downweight outliers. Notice that downweighting is performed at the expense of the number of degrees of freedom in the dataset, and may therefore result in an estimate (weighted mean) with larger errors (see Equation (3.6)) than its least-square counterpart. As confidence in the mean increases, the weighting function favours a narrower cut-off threshold from one iteration to the next.

episodic failure of the uniform source field assumption (particularly during magnetic storms). The clustering of some related errors (for example, during magnetic storms) further undermines the assumption of uncorrelated errors. Other techniques for eliminating bias arising from data points (often denoted *outliers*; see analogy shown in Figure 4.3) that are unrepresentative of the data as a whole include so-called robust processing techniques (e.g., Egbert and Booker, 1986). Robust processing involves a least-squares type algorithm, but one that is iteratively weighted to account for departures from a *Gaussian distribution* of errors. By comparing the expected versus observed distribution of residuals, a data-adaptive scheme, reliant on a function (e.g., regression *M*-estimate, Huber, 1981) containing iteratively calculated weights is implemented, with data points poorly fitting the expected robust distribution being assigned smaller weights (Egbert and Booker, 1986):

$$\sum_{i=1}^{2N} w_i r_i^2. \tag{4.21}$$

The weights, w_i, are inversely proportional to the variance $(1/\sigma_i^2)$ of the ith data point from the mean and r_i is the residual. The residual distribution is Gaussian at the centre, but is truncated and, therefore, has thicker (Laplacian) tails. If data points assigned weights below a pre-determined threshold are omitted from the stacking process, this results in a reduction of the number of degrees of freedom. From Equation (3.6), we see that a reduction in the number of degrees of freedom will have a tendency to increase the magnitudes of the error estimates for the transfer functions. However, the omission of noisy data should lead to an increase in coherence (Equation (3.8)), which tends to reduce the magnitudes of the error estimates (e.g., Equation (4.19)). A trade-off, therefore, occurs between the reduced number of degrees of freedom versus increased *coherence* (Section 3.5; Figure 3.8), and whether robust processing results in transfer functions with smaller or larger errors will depend on the amount of available data and on the degree of noise contamination.

As already mentioned (Section 4.1), the choice of length of the data window involves a trade-off between the resolution and the error of the calculated transfer function. Data outliers are more easily identified if their power is not averaged over a large number of Fourier coefficients, favouring the use of short time series windows (having, for example, 256 data points) during processing. However, if we have a sample rate of 2 s, then the longest period that can be included in such a window is of the order 500 s. Larger periods are attained through a process of *decimation*, which involves sub-sampling and low-pass filtering the data (to remove periods shorter than the Nyquist period of the decimated data). For example, by sub-sampling every fourth data point, our effective sampling rate becomes 8 s, the Nyquist period is 16 s, and a data window with 256 points now includes periods of the order 2000 s. Decimation can be performed repeatedly until the desired period range is achieved. In some decimation schemes, the digital low-pass filters do not have a steep cut-off. In our example, we might expect our decimated and undecimated datasets to produce transfer functions that overlap at 32 s. In practice, there is often a mismatch between the longest-period transfer functions calculated from the undecimated data and the shortest-period transfer functions calculated from the decimated data: the period at which the two transfer functions overlap is often shifted towards the middle of the period range spanned by the decimated data owing to inadequate low-pass filtering.

Robust processing can help to discriminate against rare, but powerful, source-field heterogeneities of short spatial scales that sometimes occur even at mid-latitudes. Robust processing methods are now widely adopted as an integral part of MT data reduction, having been shown to yield more reliable estimates of MT transfer functions than least-square processing methods in many circumstances. The flowchart shown in Figure 4.4 summarises the steps involved in processing MT data.

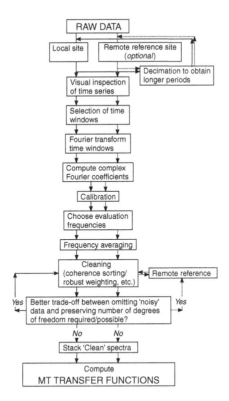

Figure 4.4 Flowchart summarising the steps involved in processing MT data.

4.3 'Upwards' and 'downwards' biased estimates

In Section 4.2 we explained how components of the impedance tensor can be isolated from inexact measurements of the electromagnetic fields using cross-power spectral densities. From Equations (4.16), we can find six pairs of simultaneous equations containing any given component of the impedance tensor, and each of these simultaneous equations can be solved to give six mathematical expressions for each impedance tensor component (Sims *et al.*, 1971). For example, (in the absence of remote reference data (Section 4.2)) the impedance tensor component Z_{ij} can be written as:

$$Z_{ij} = \frac{\langle \tilde{H}_i \tilde{E}_i^* \rangle \langle \tilde{E}_i \tilde{H}_j^* \rangle - \langle \tilde{H}_i \tilde{H}_j^* \rangle \langle \tilde{E}_i \tilde{E}_i^* \rangle}{\langle \tilde{H}_i \tilde{E}_i^* \rangle \langle \tilde{H}_j \tilde{H}_j^* \rangle - \langle \tilde{H}_i \tilde{H}_j^* \rangle \langle \tilde{H}_j \tilde{E}_i^* \rangle} \tag{4.22a}$$

or

$$Z_{ij} = \frac{\langle \tilde{H}_i \tilde{E}_j^* \rangle \langle \tilde{E}_i \tilde{H}_i^* \rangle - \langle \tilde{H}_i \tilde{H}_i^* \rangle \langle \tilde{E}_i \tilde{E}_j^* \rangle}{\langle \tilde{H}_i \tilde{E}_j^* \rangle \langle \tilde{H}_j \tilde{H}_i^* \rangle - \langle \tilde{H}_i \tilde{H}_i^* \rangle \langle \tilde{H}_j \tilde{E}_j^* \rangle} \tag{4.22b}$$

or

$$Z_{ij} = \frac{\langle \tilde{H}_i \tilde{E}_i^* \rangle \langle \tilde{E}_i \tilde{H}_i^* \rangle - \langle \tilde{H}_i \tilde{H}_i^* \rangle \langle \tilde{E}_i \tilde{E}_i^* \rangle}{\langle \tilde{H}_i \tilde{E}_i^* \rangle \langle \tilde{H}_j \tilde{H}_i^* \rangle - \langle \tilde{H}_i \tilde{H}_i^* \rangle \langle \tilde{H}_j \tilde{E}_i^* \rangle} \tag{4.22c}$$

or

$$Z_{ij} = \frac{\langle \tilde{H}_i \tilde{E}_i^* \rangle \langle \tilde{E}_i \tilde{E}_j^* \rangle - \langle \tilde{H}_i \tilde{E}_j^* \rangle \langle \tilde{E}_i \tilde{E}_i^* \rangle}{\langle \tilde{H}_i \tilde{E}_i^* \rangle \langle \tilde{H}_j \tilde{E}_j^* \rangle - \langle \tilde{H}_i \tilde{E}_j^* \rangle \langle \tilde{H}_j \tilde{E}_i^* \rangle} \tag{4.22d}$$

or

$$Z_{ij} = \frac{\langle \tilde{H}_i \tilde{H}_i^* \rangle \langle \tilde{E}_i \tilde{H}_j^* \rangle - \langle \tilde{H}_i \tilde{H}_j^* \rangle \langle \tilde{E}_i \tilde{H}_i^* \rangle}{\langle \tilde{H}_i \tilde{H}_i^* \rangle \langle \tilde{H}_j \tilde{H}_j^* \rangle - \langle \tilde{H}_i \tilde{H}_j^* \rangle \langle \tilde{H}_j \tilde{H}_i^* \rangle} \tag{4.22e}$$

or

$$Z_{ij} = \frac{\langle \tilde{H}_i \tilde{E}_j^* \rangle \langle \tilde{E}_i \tilde{H}_j^* \rangle - \langle \tilde{H}_i \tilde{H}_j^* \rangle \langle \tilde{E}_i \tilde{E}_j^* \rangle}{\langle \tilde{H}_i \tilde{E}_j^* \rangle \langle \tilde{H}_j \tilde{H}_j^* \rangle - \langle \tilde{H}_i \tilde{H}_j^* \rangle \langle \tilde{H}_j \tilde{E}_j^* \rangle}. \tag{4.22f}$$

For the 1-D case, there should be no correlation between the perpendicular electric fields, E_i and E_j, nor between the perpendicular magnetic fields H_i and H_j, whilst parallel electric and magnetic fields should also be uncorrelated, i.e., $\langle \tilde{E}_i \tilde{E}_j^* \rangle$, $\langle \tilde{E}_i \tilde{H}_i^* \rangle$, $\langle \tilde{E}_j \tilde{H}_j^* \rangle$, and $\langle \tilde{H}_i \tilde{H}_j^* \rangle$ tend to zero. As such, stable solutions cannot be determined readily from Equations (4.22a) and (4.22b).

Measured electric and magnetic fields are always likely to contain noise. Therefore, we can expand each component of the observed electromagnetic fields as the sum of the natural field plus noise. For example,

$$\left. \begin{array}{l} E_{x_{OBS}} = E_x + E_{x_{NOISE}} \\ H_{x_{OBS}} = H_x + H_{x_{NOISE}} \end{array} \right\}. \tag{4.23}$$

Assuming that the noise is incoherent (i.e., $E_{x_{NOISE}}$, $H_{x_{NOISE}}$, etc., are random and independent of the natural electromagnetic fields and of each other) then the following auto-power density spectra can be generated from Equations (4.23):

$$\begin{aligned} \langle E_{x_{OBS}} E_{x_{OBS}}^* \rangle &= \langle E_x + E_{x_{NOISE}} \rangle \langle E_x^* + E_{x_{NOISE}}^* \rangle \\ &= \langle E_x E_x^* \rangle + \underbrace{2 \langle E_x E_{x_{NOISE}} \rangle}_{=0} + \langle E_{x_{NOISE}} E_{x_{NOISE}}^* \rangle \\ &= \langle E_x E_x^* \rangle + \langle E_{x_{NOISE}} E_{x_{NOISE}}^* \rangle. \end{aligned} \tag{4.24}$$

Similarly,

$$\langle E_{y_{\mathrm{OBS}}} E^*_{y_{\mathrm{OBS}}} \rangle = \langle E_y E^*_y \rangle + \langle E_{y_{\mathrm{NOISE}}} E^*_{y_{\mathrm{NOISE}}} \rangle \tag{4.25}$$

$$\langle H_{x_{\mathrm{OBS}}} H^*_{x_{\mathrm{OBS}}} \rangle = \langle H_x H^*_x \rangle + \langle H_{x_{\mathrm{NOISE}}} H^*_{x_{\mathrm{NOISE}}} \rangle \tag{4.26}$$

$$\langle H_{y_{\mathrm{OBS}}} H^*_{y_{\mathrm{OBS}}} \rangle = \langle H_y H^*_y \rangle + \langle H_{y_{\mathrm{NOISE}}} H^*_{y_{\mathrm{NOISE}}} \rangle. \tag{4.27}$$

Meanwhile, the cross-spectral density estimates are free of noise if averaged over many samples e.g.,

$$\langle E_{x_{\mathrm{OBS}}} H^*_{y_{\mathrm{OBS}}} \rangle = \langle E_x + E_{x_{\mathrm{NOISE}}} \rangle \langle H^*_y + H^*_{y_{\mathrm{NOISE}}} \rangle$$

$$= \langle E_x H^*_y \rangle + \underbrace{\langle E_x H^*_{y_{\mathrm{NOISE}}} \rangle}_{=0} + \underbrace{\langle E_{x_{\mathrm{NOISE}}} H^*_y \rangle}_{=0} + \underbrace{\langle E_{x_{\mathrm{NOISE}}} H^*_{y_{\mathrm{NOISE}}} \rangle}_{=0}$$

$$= \langle E_x H^*_y \rangle. \tag{4.28}$$

For the simplest case of a 1-D Earth, Equations (4.22c) and (4.22d), which give stable estimates for Z_{ij} provided that the incident fields are not highly polarised, may, therefore, be expanded as:

$$Z_{ij_{\mathrm{OBS}}} = Z_{ij} \left(1 + \frac{\langle E_{i_{\mathrm{NOISE}}} E^*_{i_{\mathrm{NOISE}}} \rangle}{\langle E_i E^*_i \rangle} \right). \tag{4.29}$$

Equations (4.22e) and (4.22f) also give stable estimates for Z_{ij} provided that the incident fields are not highly polarised, and may be expanded as:

$$Z_{ij_{\mathrm{OBS}}} = Z_{ij} \left(1 + \frac{\langle H_{j_{\mathrm{NOISE}}} H^*_{j_{\mathrm{NOISE}}} \rangle}{\langle H_j H^*_j \rangle} \right)^{-1}. \tag{4.30}$$

From Equations (4.29) and (4.30), we see that estimates of Z_{ij} based on Equations (4.22c) and (4.22d) are biased upwards by incoherent noise in the measured electric fields, whilst estimates of Z_{ij} based on Equations (4.22e) and (4.22f) are biased downwards by incoherent noise in the measured magnetic fields. The degree of bias depends on the signal-to-noise ratio – the more power present in the noise spectra, the greater the bias to Z_{ij}.

In summary, of the six mathematical expressions for Z_{ij}, two are biased downward by any random noise present in the magnetic field components, but are unaffected by random noise superimposed on the electric signal, two are biased upwards by random noise imposed on the electric field components, but are unaffected by any random noise present in the magnetic field components, and the remaining two are unstable and should be avoided.

One method of reducing the effects of this type of biasing of impedance tensor estimates is to incorporate remote reference data (Section 4.2). Another possibility is to calculate a weighted average (Z_{ij}^{AV}) of the up (Z_{ij}^{U}) and down (Z_{ij}^{D}) biased estimates (e.g., Jones 1977) normalised by the squares of their associated errors (ΔZ_{ij}^{U} and ΔZ_{ij}^{D}):

$$Z_{ij}^{AV} = \left(\frac{Z_{ij}^{U}}{\left(\Delta Z_{ij}^{U}\right)^2} + \frac{Z_{ij}^{D}}{\left(\Delta Z_{ij}^{D}\right)^2} \right) \left(\frac{1}{\left(\Delta Z_{ij}^{U}\right)^2} + \frac{1}{\left(\Delta Z_{ij}^{D}\right)^2} \right)^{-1}. \qquad (4.31)$$

4.4 Displaying the data and other deceptions

Some guidelines for data presentation are given in Hobbs (1992). Presentation of single-site transfer functions is generally via Cartesian graphs of *apparent resistivities* and *impedance phases* (e.g., Figure 4.5) plotted as a function of $\log_{10} T$ (along the abscissa). For apparent resistivity graphs, the ordinate is $\log_{10} \rho_a$, with values increasing upwards. For impedance phase graphs, the ordinate is linear with

Figure 4.5 Apparent resistivity and impedance phase as a function of period for an MT site (NGU) placed a few hundred metres from the Nguramani escarpment, which flanks the Rift Valley in southernmost Kenya (see Figure 5.13(a)).

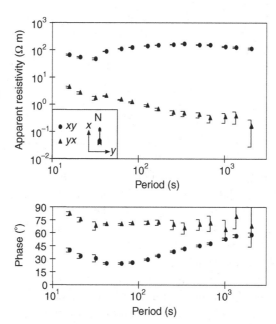

(a)

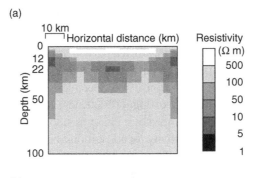

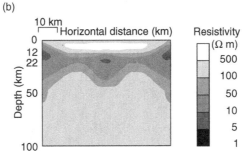

(b)

Figure 4.6 (a) 2-D block model output by least-structure inversion (see Chapter 7). (b) Contour plot derived from (a). Compare also Figure 7.4 (c).

units of degrees (e.g., 0–90°). When plotting multi-site data as coloured *pseudosections* (see Section 4.5) or 2-D models, a relative scheme has been recommended, in which red represents low apparent resistivity (high impedance phases) and blue represents high apparent resistivity (low impedance phases).

MT data resolve conductance and conductivity gradients, rather than conductivity. If least-structure models (see, for example, Chapter 7) are presented as contour plots, then the steps in the colour or grey scale can subjectively influence how the model is interpreted. Figure 4.6 (a) shows discrete resistivity values obtained from 2-D inversion, whilst Figure 4.6 (b) is contoured from the discrete resistivities in Figure 4.6 (a). An irregularly shaped anomaly is present in both of the representations shown in Figure 4.6, but how we define its depth and dimensions depends on what resistivity value or contour we choose to define the transition from resistive host to conductive anomaly. Compare Figure 4.6 (b) with Figure 7.4 (c), where the same model has been contoured using a different grey scale. Does one of the grey scales hint at a larger, more conductive anomaly than the other? This kind of subjective manipulation of data and models tends to be accentuated when a colour scale is used. Try photocopying Figure 4.6 and colouring it in according to different colour scales.

4.5 Through data interpolation to pseudosections

A common way of displaying multi-site, multi-frequency data is as a pseudosection, in which a particular parameter (e.g., E-polarisation or B-polarisation apparent resistivity or impedance phase) is interpolated both as a function of distance along a profile and as a function of $\log_{10}$ (period), or $\log_{10}$ (frequency) to produce a contour plot (e.g., Figure 4.7). Impedance phase pseudosections are often favoured over apparent resistivity pseudosections, because impedance phases are free of static shift. There are various mathematical methods of interpolating data (see, for example, Akima, 1978), and subjective choices that can be made regarding contour intervals, and experimenting with these will alter the fine structure of the pseudosection produced.

Figure 4.7 'B-polarisation' apparent resistivity pseudosection derived from a profile of MT sites traversing the southernmost part of the Kenya Rift (see Figure 5.13(a)). The false 'intrusion' is an artefact generated by a near-surface conductivity discontinuity between the Rift Valley and its western boundary. Dashed line indicates period extent of the data.

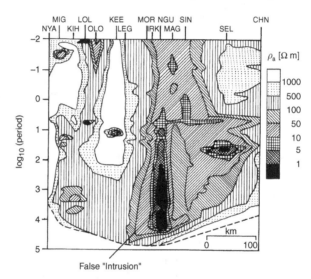

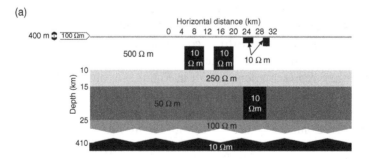

Figure 4.8 (a) 2-D model. (b) – (f) E-polarisation impedance phase pseudosections synthesised from (a) with site spacings of: (b) 1 km, (c) 2 km, (d) 4 km, (e) 8 km, and (f) 16 km. (g) – (k) B-polarisation impedance phase pseudosections synthesised from (a) with site spacings of: (g) 1 km, (h) 2 km, (i) 4 km, (j) 8 km, and (k) 16 km. All pseudosections are interpolated from 17 periods that increase in the sequence 0.016 s, 0.032 s, 0.064 s, ..., 1048 s (i.e., ~ three periods per decade). Notice that whereas the E-polarisation has better vertical resolution, the B-polarisation has better lateral resolution.

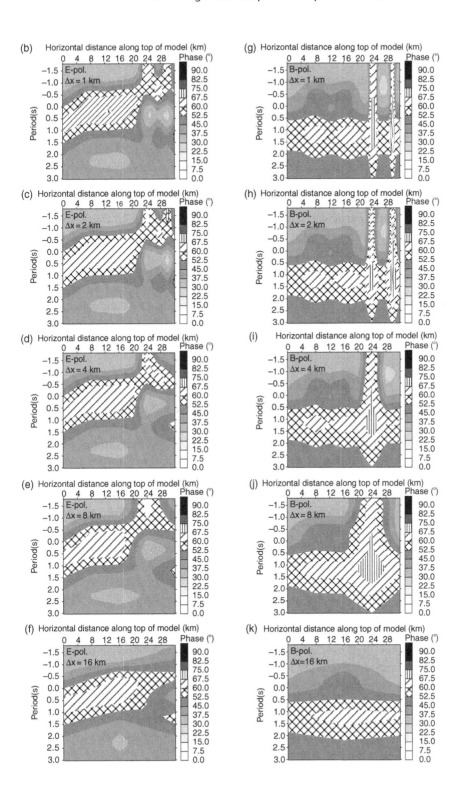

Figure 4.8 compares impedance phase pseudosections generated by a simple 2-D conductivity model for data synthesised at different site densities. Notice that the **B**-polarisation affords better lateral resolution, whereas the **E**-polarisation affords better vertical resolution. The *spatial aliasing* (Section 3.2) of lateral conductivity variations increases with increasing site spacing, and there is an inherent lack of lateral resolution for structures at depth. What conductivity structures are reliably reconstructed for a given site density?

When using pseudosections to make a preliminary interpretation of data, the number of periods/sites supporting the presence of a particular structure should be considered. In this respect it is sometimes helpful to overlay the discrete data points from which the pseudosection has been produced.

In relatively simple environments, pseudosections can provide a useful visual aid by highlighting conductors prior to modelling. However, in the presence of multi-dimensional conductivity structures, pseudosections can be highly deceptive, as demonstrated in the next section.

4.6 Pseudosections versus Cartesian graphs

As discussed in Section 2.6, changes in conductivity across a boundary generate discontinuous electric fields in the direction perpendicular to the boundary, as a consequence of conservation of current. This can lead to highly anomalous apparent resistivities and impedance phases across a wide period range. For example, Figure 4.5 shows Cartesian graphs of apparent resistivity and impedance phase as a function of period, obtained from an MT site placed a few hundred metres from a sharp conductor–insulator boundary, that extends to only ~3 km depth. '**B**-polarisation' (see Section 5.6 to understand why we enclose '**B**-polarisation' in inverted commas) apparent resistivity pseudosections from a profile across the boundary are shown in Figure 4.7. The site from which the highly anomalous responses shown in Cartesian form in Figure 4.5 were acquired lies approximately halfway along the profile represented in Figure 4.7. In pseudosection, these data appear to indicate a conductive body at long periods, giving the false impression of a conductive intrusion at depth.

Another disadvantage of presenting data in the form of pseudosections is a tendency on the part of the observer to overlook errors in the data when interpreting them. On the other hand, data from a large number of sites can be more easily digested when presented as a pseudosection than as a large number of site-specific Cartesian graphs.

Chapter 5
Dimensionality and distortion

So you managed to estimate some transfer functions for your first site, and the penetration depth for the period one hour is 3500 km! This is in the outer core of the Earth, and although we learned in Chapter 2 that the penetration depth depends on both period and conductivity, 3500 km is certainly deeper than we should expect to probe in any tectonic environment. What is going on? The transfer functions are affected by 'static shift', caused by small-scale conductivity contrasts. The 'shift' refers to the apparent resistivity curve, which is in fact shifted (in your case: towards a larger apparent resistivity, but a shift in the opposite direction can also occur). The shift is called 'static', because the underlying physical principle, the conservation of current at conductivity discontinuities, is not a time-dependent process like, for example, induction. Therefore, the phenomenon of static shift does not affect the phase of the transfer function. However, in complex conductivity distributions, frequency-independent distortion effects, commonly referred to as 'galvanic effects', can result in mixing of the phases belonging to transfer functions describing two horizontal directions (e.g., parallel and perpendicular to a vertical boundary). Sophisticated techniques for recovering the undistorted transfer functions have been developed. Three mistakes frequently occur: not using any of these techniques where they are necessary, over-using them in simple cases, and ignoring indications that the assumptions underlying these techniques have broken down (due, for example, to poor data quality, or three-dimensional (3-D) induction). We present a series of general models of increasing complexity, and try to advise the right level of 'distortion removal'. In some cases, more than one strike direction influences the same data. A good meeting point of general geoscience and

MT are the classes of models that incorporate small-scale conductivity contrasts, electrical anisotropy and shear zones with a particular strike direction.

5.1 The discontinuity revisited: the concept of static shift

Figure 5.1(d) shows *apparent resistivities* and *impedance phases* calculated across a discontinuity (Figure 5.1(a)). What are the *penetration depths* for the two polarisations of the data at a period of one hour? Do these penetration depths seem realistic? A penetration depth of ~1100 km (for the **B**-*polarisation*) would lie within the Earth's lower mantle, and considering that rapid attenuation of electromagnetic fields can be expected in the mid-mantle transition zone (410–660 km depth range), the penetration depth calculated from the **B**-polarisation appears to be exaggerated. In contrast, the penetration depth of ~300 km suggested by the **E**-*polarisation* lies shallower than the mid-mantle transition region, and the idea that

Figure 5.1 (a) 2-D model consisting of two 5-km-thick quarter-spaces having resistivities of 10 Ω m and 1000 Ω m, respectively underlain by a 100 Ω m half-space, and terminating in a 5 Ω m half-space extending downwards from a depth of 410 km. (b) 1-D model in which the 5-km-thick quarter-spaces shown in (a) are replaced by a 1000 Ω m half-space (i.e., 10 Ω m quarter-space removed). (c) 1-D model in which the 5-km-thick quarter-spaces shown in (a) are replaced by a 10 Ω m *half-space* (i.e., 1000 Ω m quarter-space removed). (d) Comparison between the apparent resistivities and impedance phases synthesised for a site placed on the resistive quarter-space ~300 m from the boundary shown in (a) and the 1-D apparent resistivities and impedance phases synthesised from (b) and (c). At long periods, the 1-D and 2-D E-polarisation apparent resistivities converge to 100 Ω m, which is the resistivity of the *half-space* extending from 5–410 km, whereas the B-polarisation apparent resistivities are shifted by a constant factor of ~13 for periods longer than ~10 s.

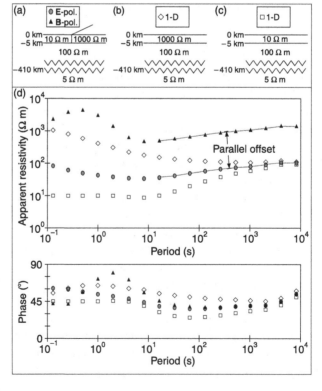

the electromagnetic fields are confined within the upper mantle is supported by the forms of the apparent resistivity and phase curves, which show no evidence for an highly conductive layer (below 410 km) being penetrated. In fact, the **E**-polarisation apparent resistivities converge to those synthesised from layered-Earth models (Figures 5.1(b) and 5.1(c)) at long periods.

The data shown in Figure 5.1 display the phenomenon of *static shift*. Static shift can be caused by any multi-dimensional conductivity contrasts having depths and dimensions less than the true penetration depth of electromagnetic fields. Conductivity discontinuities cause local distortion of the amplitudes of electric fields as a result of conservation of electric charge, hence causing impedance magnitudes to be enhanced or diminished by real scaling factors. Electric charges were originally neglected in the derivation of the *diffusion equation* (Section 2.3) because we assumed $\nabla \cdot \mathbf{E} = 0$. In those cases where a current crosses a discontinuity (this situation was previously described in Equation (2.11) and in more detail in Section 2.6) charges build up along the discontinuity. The resulting shift in apparent resistivity curves is referred to as 'static' because, unlike induction, conservation of current is not a time-dependent process. The time-independent nature of static shift means that there is no impedance phase associated with the phenomenon. Indeed, the presence of static shift is most easily identifiable in measured data in which apparent resistivities are shifted relative to each other, but impedance phases lie together. As a result of static shift, apparent resistivity curves are shifted by a constant, real scaling factor, and therefore preserve the same shape as unshifted apparent resistivity curves in the period range where static shift occurs. Static shifts are generally more prevalent in highly resistive environments, where small-scale conductivity heterogeneities have a more significant effect on electric fields.

Non-inductive responses are also commonly referred to as *galvanic effects*. Some types of galvanic effects that arise in complex geoelectric environments do affect impedance phases, by causing mixing of differently polarised data. Static shift is a subset of these types of galvanic effects.

5.2 Rotating the impedance tensor

In a layered Earth, the impedance in all directions can be calculated (from the bivariate regression expressed in Equation (4.17)) simply by measuring the electric field variation in one direction and the magnetic field variation in the perpendicular direction. We could

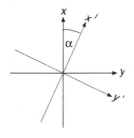

Figure 5.2. A rotation through angle α maps x to x' and y to y'.

test the hypothesis of a layered Earth either (i) by performing the measurements in different co-ordinate frames – (x, y) and (x', y') – and comparing the elements of the impedance tensors, $\underline{\underline{Z}}$ and $\underline{\underline{Z}}'$, respectively, or (ii) by applying a mathematical rotation to the impedance tensor estimated from data measured in a fixed co-ordinate frame. Theoretically, we can simulate a field site setup with sensors oriented in any direction, α (Figure 5.2), via a mathematical rotation involving matrix multiplication of the measured fields (or impedance tensor) with a rotation matrix, $\underline{\underline{\beta}}_\alpha$:

$$\left.\begin{array}{l} \mathbf{E}' = \underline{\underline{\beta}}_\alpha \, \mathbf{E} \\ \mathbf{B}' = \underline{\underline{\beta}}_\alpha \, \mathbf{B} \end{array}\right\} \therefore \underline{\underline{\beta}}_\alpha \, \mathbf{E} = \underline{\underline{Z}}'\underline{\underline{\beta}}_\alpha \, (\mathbf{B}/\mu_0) \Rightarrow \underline{\underline{Z}}' = \underline{\underline{\beta}}_\alpha \, \underline{\underline{Z}}\underline{\underline{\beta}}_\alpha^{\mathrm{T}} \tag{5.1}$$

where

$$\underline{\underline{\beta}}_\alpha = \begin{pmatrix} \cos\alpha & \sin\alpha \\ -\sin\alpha & \cos\alpha \end{pmatrix} \tag{5.2}$$

and superscript T denotes the transpose of $\underline{\underline{\beta}}_\alpha$,

$$\underline{\underline{\beta}}_\alpha^{\mathrm{T}} = \begin{pmatrix} \cos\alpha & -\sin\alpha \\ \sin\alpha & \cos\alpha \end{pmatrix}. \tag{5.3}$$

Equation (5.1) can be expanded as,

$$\left.\begin{array}{l} Z'_{xx} = Z_{xx}\cos^2\alpha + (Z_{xy} + Z_{yx})\sin\alpha\cos\alpha + Z_{yy}\sin^2\alpha \\ Z'_{xy} = Z_{xy}\cos^2\alpha + (Z_{xx} + Z_{yy})\sin\alpha\cos\alpha - Z_{yx}\sin^2\alpha \\ Z'_{yx} = Z_{yx}\cos^2\alpha + (Z_{yy} - Z_{xx})\sin\alpha\cos\alpha - Z_{xy}\sin^2\alpha \\ Z'_{yy} = Z_{yy}\cos^2\alpha + (Z_{yx} - Z_{xy})\sin\alpha\cos\alpha - Z_{xx}\sin^2\alpha \end{array}\right\}. \tag{5.4}$$

Or, more elegantly,

$$\left.\begin{array}{l} Z'_{xx} = S_1 + S_2\sin\alpha\cos\alpha \\ Z'_{xy} = D_2 + S_1\sin\alpha\cos\alpha \\ Z'_{yx} = D_2 - D_1\sin\alpha\cos\alpha \\ Z'_{yy} = -D_1 - D_2\sin\alpha\cos\alpha \end{array}\right\}, \tag{5.5}$$

where S_1, S_2, D_1 and D_2 are the *modified impedances* (Vozoff, 1972):

$$\left.\begin{array}{ll} S_1 = Z_{xx} + Z_{yy} & S_2 = Z_{xy} + Z_{yx} \\ D_1 = Z_{xx} - Z_{yy} & D_2 = Z_{xy} - Z_{yx} \end{array}\right\}. \tag{5.6}$$

The rotated modified impedances are simply

$$S'_1 = Z'_{xx} + Z'_{yy} = Z_{xx} + Z_{yy} = S_1$$

$$\begin{aligned} D'_1 &= Z'_{xx} - Z'_{yy} \\ &= (\cos^2\alpha - \sin^2\alpha)(Z_{xx} - Z_{yy}) + 2\cos\alpha\sin\alpha(Z_{xy} + Z_{yx}) \\ &= \cos^2\alpha \, D_1 + \sin^2\alpha \, S_2 \end{aligned} \tag{5.7a}$$

and, similarly,

$$S_2' = \cos^2\alpha \, S_2 - \sin^2\alpha \, D_1 \quad \text{and} \quad D_2' = D_2. \qquad (5.7b)$$

Hence, S_1 and D_2 are *rotationally invariant*.

5.3 A parade of general models and their misfit measures

In the 1960s, 70s and 80s, the initial 1-D (layered-Earth) MT model was superseded by a series of general conductivity models which allowed for more structure than a 1-D model, but not for complete three-dimensionality. The first of these general models was suggested by Cantwell (1960), who replaced scalar impedance with a rank 2 impedance tensor $\underline{Z}$ that links the two-component horizontal electric and magnetic variational fields **E** and **H**:

$$\begin{pmatrix} E_x \\ E_y \end{pmatrix} = \begin{pmatrix} Z_{xx} & Z_{xy} \\ Z_{yx} & Z_{yy} \end{pmatrix} \begin{pmatrix} H_x \\ H_y \end{pmatrix}. \qquad (5.8)$$

In the layered-Earth model considered by Tikhonov (1950, reprinted 1986), Cagniard (1953) and Wait (1954) the impedance tensor would be:

$$\begin{pmatrix} E_x \\ E_y \end{pmatrix} = \begin{pmatrix} 0 & Z_n \\ -Z_n & 0 \end{pmatrix} \begin{pmatrix} H_x \\ H_y \end{pmatrix} \qquad (5.9)$$

in any co-ordinate system (because parallel components of **E** and **H** should be uncorrelated). A rotationally invariant dimensionless misfit measure that can be used to check whether the layered-Earth model described by Equation (5.9) is appropriate for measured data is:

$$\sum = (D_1^2 + S_2^2)/D_2^2. \qquad (5.10)$$

If Σ is significantly larger than the relative error $\Delta D_2/|D_2|$, then the data require a more complex model.

In the case of a 2-D conductivity structure, (e.g., two quarter-spaces with different layered structures (Figure 2.5), or a dyke) the impedance tensor can – in an appropriate co-ordinate system – be reduced to

$$\begin{pmatrix} E_x \\ E_y \end{pmatrix} = \begin{pmatrix} 0 & Z_{xy} \\ Z_{yx} & 0 \end{pmatrix} \begin{pmatrix} H_x \\ H_y \end{pmatrix} \qquad (5.11)$$

with Z_{xy} and Z_{yx} being the different impedances of two decoupled systems of equations describing induction with electric fields parallel and perpendicular, respectively, to the *electromagnetic strike* of the 2-D structure (as, for example, described in Section 2.6).

Measured data rarely have zero diagonal impedance tensor elements in any co-ordinate system. This may be a result of:

(i) data errors imposed on a truly 1-D or 2-D inductive response;
(ii) coupling of the regional 1-D or 2-D inductive response with localised, small-scale (relative to the scale of observation), 3-D heterogeneities that cause non-inductive distortion effects; or
(iii) 3-D induction effects.

If departures from the ideal situation (in which a rotation angle can be found that completely annuls the diagonal elements of the impedance tensor) arise owing to extraneous noise, then it may be appropriate to compute an electromagnetic strike as the angle, α, in Equation 5.4 that maximises the off-diagonal elements and minimises the diagonal elements of the impedance tensor (Swift (1967, reprinted 1986)). In practice, there are numerous conditions that may be applied in order to determine α, and thus retrieve the two principal, off-diagonal impedances for this simple 2-D case: e.g., maximising $|Z_{xy_{2D}}|$ or $|Z_{yx_{2D}}|$, minimising $|Z_{xx_{2D}}|$ or $|Z_{yy_{2D}}|$, maximising $|Z_{xy_{2D}} + Z_{yx_{2D}}|$, minimising $|Z_{xx_{2D}} + Z_{yy_{2D}}|$, maximising $|Z_{xy_{2D}}|^2 + |Z_{yx_{2D}}|^2$, or minimising $|Z_{xx_{2D}}|^2 + |Z_{yy_{2D}}|^2$. For a truly 2-D structure, all of these criteria should yield the same principal directions (although, in practice, they often do not!)

For example, the angle, α, at which $|Z_{xx}(\alpha)|^2 + |Z_{yy}(\alpha)|^2$ is minimised is obtained by satisfying the condition that

$$\frac{\partial}{\partial \alpha}\left(|Z_{xx}(\alpha)|^2 + |Z_{yy}(\alpha)|^2\right) = 0. \tag{5.12}$$

The condition expressed in Equation (5.12) is satisfied for

$$\tan 4\alpha = \frac{2\mathrm{Re}(S_2' D_1')}{|D_1'|^2 - |S_2'|^2}. \tag{5.13}$$

The second derivative of $|Z_{xx}(\alpha)|^2 + |Z_{yy}(\alpha)|^2$ also needs to be considered, in order to confirm that the diagonal components are **minimised** by a rotation through angle α. Otherwise, a maximum is found and

$$\alpha \to \alpha + 45° \tag{5.14}$$

(Swift, 1967, reprinted 1986; Vozoff, 1972).

The computed strike direction contains a 90° ambiguity, because rotation by 90° only exchanges the location of the two principal impedance tensor elements within the tensor:

$$\underline{\beta}_{90}\begin{pmatrix} 0 & Z_{xy} \\ Z_{yx} & 0 \end{pmatrix}\underline{\beta}_{90}^T = \begin{pmatrix} 0 & 1 \\ -1 & 0 \end{pmatrix}\begin{pmatrix} 0 & Z_{xy} \\ Z_{xy} & 0 \end{pmatrix}\begin{pmatrix} 0 & -1 \\ 1 & 0 \end{pmatrix}$$
$$= \begin{pmatrix} 0 & -Z_{yx} \\ -Z_{xy} & 0 \end{pmatrix}. \tag{5.15}$$

Therefore, electromagnetic strikes of α and $\alpha \pm 90°$ cannot be separated using a purely mathematical model.

The early availability of numerical solutions for induction in 2-D structures (e.g., Jones and Pascoe, 1971) contributed to the widespread use of the *Swift model*. Swift (1967, reprinted 1986) also provided a rotationally invariant misfit parameter referred to as *Swift skew*, κ,:

$$\kappa = |S_1|/|D_2|, \tag{5.16}$$

which provides an 'adhoc' indication as to the appropriateness of applying the Swift model to measured data (see also Vozoff, 1972). The errors of the *Swift strike* and Swift skew are

$$\Delta(\tan\alpha) = \left\{ \left[\left(\frac{2\Delta S_2 D_1}{|D_1|^2 - |S_2|^2} \right)^2 + \left(\frac{2\Delta D_1 S_2}{|D_1|^2 - |S_2|^2} \right)^2 \right. \right.$$
$$\left. + \left(\frac{2\Re(S_2 D_1)}{\left(|D_1|^2 - |S_2|^2\right)^2} \right)^2 \right] \tag{5.17}$$
$$\left. \left((\Delta D_1 D_1)^2 + (\Delta S_2 S_2)^2 \right) \right\}^{1/2}$$

$$\Delta\kappa = \left[\left(\frac{\Delta S_1}{D_2} \right)^2 + \left(\frac{\Delta D_2 S_1}{D_2^2} \right)^2 \right]^{1/2}, \tag{5.18}$$

where the error functions denoted by Δ are real.

If distortions of the type discussed in (ii) above are present, then the electromagnetic strike determined from the Swift method is unlikely to lead to satisfactory unmixing of the principal imped-ances, and a more complex model will be required.

In Section 2.9, we discussed how the dimensionality of a con-ductivity anomaly depends on the scale at which we observe it (Figure 2.11). When the electromagnetic *skin depth* becomes several times greater than the dimensions of the anomaly, its electromagnetic

Figure 5.3 A near-surface 3-D conductor with resistivity ρ_2 appears 1-D for MT sounding periods that are sufficiently short to be contained within the conductor, but as the sounding period lengthens, edges I and II are detected and the MT response is one of 2-D induction. Then, as the inductive scale length increases sufficiently to image edges I–IV, the finite strike of the conductor is revealed by a 3-D inductive response. Owing to charges at the boundaries, galvanic effects are also produced, and at periods for which the electromagnetic skin depth is significantly greater than the dimensions of the near-surface inhomogeneity, the galvanic response dominates over the inductive response.

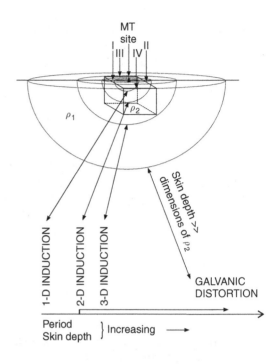

response becomes inductively weak, but it continues to have a non-inductive response (commonly termed 'galvanic'). In an early contribution of 3-D modelling to our understanding of non-inductive distortion, Wannamaker *et al.* (1984a) demonstrated this type of behaviour arising from a 3-D body embedded in a layered Earth. Electromagnetic data containing galvanic effects can often be described by a *superimposition* or *decomposition model* in which the data are decomposed into a non-inductive response owing to multi-dimensional heterogeneities with dimensions significantly less than the *inductive scale length* of the data (often described as *local*), and a response owing to an underlying 1-D or 2-D structure (often described as *regional*; Figure 5.3). In such cases, determining the electromagnetic strike involves *decomposing* the measured impedance tensor into matrices representing the inductive and non-inductive parts. The inductive part is contained in a tensor composed of components that have both magnitude and phase (i.e., its components are complex), whereas the non-inductive part exhibits DC behaviour only, and is described by a *distortion tensor,* the components of which must be **real** and therefore **frequency independent**.

Larsen (1975) considered the superposition of a regional 1-D layered-Earth model and a small-scale structure of anomalous conductance at the Earth's surface (Figure 5.4). If the size of the anomaly is small compared to the penetration depth, p, of the

(a)

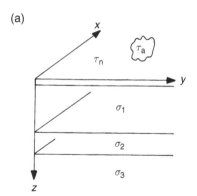

(b)

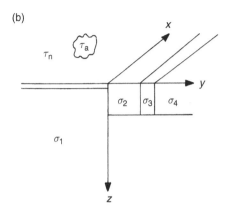

Figure 5.4 (a) Larsen's (1975) superimposition model (Equation (5.19)), in which a small-scale, near-surface conductance anomaly, τ_a, scatters electric fields, distorting the measured impedance away from that of the regional 1-D layered model in which τ_a is embedded. (b) Bahr's (1988) superimposition model (Equation (5.22)) in which a small-scale, near-surface conductance anomaly, τ_a, scatters electric fields, distorting the measured impedance away from that of the regional 2-D model in which τ_a is embedded. (Subsequent testing of the superimposition hypothesis on measured data has demonstrated that (where the hypothesis is applicable) the distorting body is not necessarily surficial).

electromagnetic field, then the impedance tensor associated with Larsen's general model can be decomposed as:

$$\begin{pmatrix} E_x \\ E_y \end{pmatrix} = \underline{\underline{C}} \begin{pmatrix} 0 & Z \\ -Z & 0 \end{pmatrix} \begin{pmatrix} H_x \\ H_y \end{pmatrix} = \begin{pmatrix} c_{11} & c_{12} \\ c_{21} & c_{22} \end{pmatrix} \begin{pmatrix} 0 & Z \\ -Z & 0 \end{pmatrix} \begin{pmatrix} H_x \\ H_y \end{pmatrix}, \quad (5.19)$$

where Z is the regional impedance of Cagniard's (1953) 1-D model and $\underline{\underline{C}}$ is a real distortion tensor, which describes the galvanic (rather than inductive) action of the local scatterer on the electric field. Equation (5.19) contains six *degrees of freedom* – four real distortion parameters and a complex impedance – whereas measured data that can be explained by this model provide only five degrees of freedom – four amplitudes and one common phase. Therefore, the absolute amplitude of the impedance cannot be separated from the distortion parameters. This is equivalent to stating that *static shift factors* cannot be mathematically determined from a measured impedance. However, moderate distortion of synthetic and measured impedance amplitudes can be analysed with simple 3-D models (Park, 1985).

Can measured data be explained with Larsen's (1975) model? If so, then all elements of the measured impedance should have the same *phase*. The difference between the phases of two *complex numbers* x and y can be determined using the commutator:

$$[x, \ y] = \text{Im}(y \ x^*)$$
$$= \text{Re}x\text{Im}y - \text{Re}y\text{Im}x.$$

(5.20)

Hence, a rotationally invariant dimensionless misfit measure for the Larsen (1975) model has been proposed as:

$$\mu = (| \ [D_1, \ S_2] \ | + | \ [S_1, \ D_2] \ |)^{1/2}/|D_2|.$$

(5.21)

For some time it was supposed that the two complementary models of Swift (Equation (5.11)) and Larsen (Equation (5.19)) would be adequate to describe measured impedances. That is to say, it was expected that a co-ordinate system should exist for which either (i) the off-diagonal components (Z_{xy} and Z_{yx}) of the impedance tensor would have different phases (due to different conductivity structures along and across strike), and the diagonal components (Z_{xx} and Z_{yy}) would be negligible; or (ii) all elements of the impedance tensor would exhibit the same phase, but the diagonal components would be non-zero. However, as demonstrated by Ranganayaki (1984), MT phases can depend strongly on the direction in which the electric field is measured, and the existence of a large class of measured impedances having both non-vanishing diagonal components and two different phases, such that they concurred with neither the Swift model nor the Larsen model, led Bahr (1988) to propose a more complete superimposition (decomposition) model. In this model, multi-dimensional heterogeneities with dimensions significantly less than the inductive scale length of the data are superposed on a regional 2-D structure, and the data are decomposed into a 'local', non-inductive response (galvanic), and a 'regional', inductive response. For data aligned in the (x', y') co-ordinate system of the regional 2-D structure, the impedance tensor can then be expanded as:

$$\begin{pmatrix} E_x \\ E_y \end{pmatrix} = \underline{\underline{C}} \begin{pmatrix} 0 & Z_{n,x'y'} \\ Z_{n,y'x'} & 0 \end{pmatrix} \begin{pmatrix} H_x \\ H_y \end{pmatrix}$$
$$= \begin{pmatrix} -c_{12}Z_{n,y'x'} & c_{11}Z_{n,x'y'} \\ -c_{22}Z_{n,y'x'} & c_{21}Z_{n,x'y'} \end{pmatrix} \begin{pmatrix} H_x \\ H_y \end{pmatrix}.$$

(5.22)

Within each column, only one phase occurs, because the assumption of galvanic distortion requires that the elements of the distortion tensor, $\underline{\underline{C}}$, must be real and frequency independent. In an arbitrary co-ordinate system, however, the phases of the two regional impedances $Z_{n,xy}$ and $Z_{n,yx}$ will be mixed, because in this case the impedance tensor elements are linear combinations of $Z_{n,xy}$ and $Z_{n,yx}$. In an arbitrary co-ordinate system, we have:

$$\underline{\underline{Z}} = \underline{\beta}_\alpha \, \underline{\underline{C}} \, \underline{\underline{Z}}_{2D} \underline{\beta}_\alpha^T. \tag{5.23}$$

The condition that, in the co-ordinate system of the regional strike, the tensor elements in the columns of the tensor should have the same phase, i.e,

$$\frac{\text{Re}(Z_{xx})}{\text{Im}(Z_{xx})} = \frac{\text{Re}(Z_{yx})}{\text{Im}(Z_{yx})} \Rightarrow \frac{\text{Re}(Z_{xx})}{\text{Re}(Z_{yx})} = \frac{\text{Im}(Z_{xx})}{\text{Im}(Z_{yx})} \tag{5.24}$$

leads to an equation for the rotation angle, α:

$$- A \sin(2\alpha) + B \cos(2\alpha) + C = 0,$$

where

$$
\begin{aligned}
A &= [S_1, D_1] + [S_2, D_2] \\
B &= [S_1, S_2] - [D_1, D_2] \\
C &= [D_1, S_2] - [S_1, D_2].
\end{aligned}
\tag{5.25}
$$

The two solutions:

$$\tan \alpha_{1,2} = \pm \left\{ (B+C)/(B-C) + [A/(B-C)]^2 \right\}^{1/2} - A/(B-C) \tag{5.26}$$

lead to a co-ordinate frame where the same phase condition is fulfilled either in the left or in the right column of the impedance tensor. Only if the commutator C is negligible, does a unified co-ordinate frame exist for which the same phase condition is fulfilled in both columns of the impedance tensor. In this case, the rotation angle, α, is found from:

$$\tan \alpha = \pm \sqrt{1 + \left(\frac{A}{B}\right)^2} - \frac{A}{B} \quad \text{or} \quad \tan(2\alpha) = B/A. \tag{5.27}$$

The electromagnetic strike that is recovered using the Bahr model is often referred to as the *phase-sensitive strike*.

For cases in which no rotation angle can be found for which the phases in the respective columns of the impedance tensor are equal, Bahr (1991) proposes minimising the phase difference δ between the elements of a given column such that:

$$\underline{\underline{Z}} = \underline{\beta} \begin{pmatrix} -c_{12} Z_{yx_r} e^{i\delta} & c_{11} Z_{xy_r} \\ -c_{22} Z_{yx_r} & c_{21} Z_{xy_r} e^{-i\delta} \end{pmatrix} \underline{\beta}^T. \tag{5.28}$$

In this case,

$$\frac{\text{Re}(Z_{x'x'})\cos\delta + \text{Im}(Z_{x'x'})\sin\delta}{\text{Re}(Z_{y'x'})} = \frac{-\text{Re}(Z_{x'x'})\sin\delta + \text{Im}(Z_{x'x'})\cos\delta}{\text{Re}(Z_{y'x'})}.$$

(5.29)

Equation (5.29) can be solved for α and δ (Bahr, 1991; Prácser and Szarka, 1999[6]) to yield:

$$\tan(2\alpha_{1,2}) = \frac{1}{2}\frac{(B_1 A_2 + A_1 B_2 + C_1 E_2)}{(A_1 A_2 - C_1 C_2 + C_1 F_2)}$$
$$\pm \left[\frac{1}{4}\frac{(B_1 A_2 + A_1 B_2 + C_1 E_2)^2}{(A_1 A_2 - C_1 C_2 + C_1 F_2)^2} - \frac{(B_1 B_2 - C_1 C_2)}{(A_1 A_2 - C_1 C_2 + C_1 F_2)}\right]^{1/2},$$

(5.30)

where

$$A = A_1 + A_2 = ([S_1, D_1] + [S_2, D_2])\cos\delta + (\{S_1, D_1\} + \{S_2, D_2\})\sin\delta$$
$$B = B_1 + B_2 = ([S_1, S_2] - [D_1, D_2])\cos\delta + (\{S_1, S_2\} - \{D_1, D_2\})\sin\delta$$
$$C = C_1 + C_2 = ([D_1, S_2] - [S_1, D_2])\cos\delta + (\{D_1, S_2\} - \{S_1, D_2\})\sin\delta$$
$$E = E_2 = (\{S_2, S_2\} - \{D_1, D_1\})\sin\delta, \quad F_2 = 2\{D_1, S_2\}\sin\delta,$$

and

$$\delta = \tan^{-1}\left[\frac{C_1}{(-A_2\sin(2\alpha) + B_2\cos(2\alpha))}\right],$$

(5.31)

where the commutator $[x, y]$ is defined by Equation (5.20) and $\{x, y\} = \text{Re}(yx^*)$. Bahr (1991) refers to this extension of the superimposition model as the *delta (δ) technique*.

Bahr (1988) proposed a rotationally invariant parameter termed *phase-sensitive skew* as an ad hoc measure of the extent to which an impedance tensor can be described by Equations (5.22) or (5.28):

$$\eta = \sqrt{C}/|D_2| = (|\,[D_1, S_2] - [S_1, D_2]\,|)^{1/2}/|D_2|.$$

(5.32)

For $\eta < 0.1$, Bahr suggested that Equation (5.22) is the appropriate model, whereas for $0.1 < \eta < 0.3$ Equation (5.28) might be more appropriate. However, see our warnings concerning skew in Section 5.7.

Another technique that is routinely used to solve for the electromagnetic strike using the decomposition hypothesis is an inverse technique proposed by Groom and Bailey (1989). Central to the decomposition hypothesis, whether solved using the Bahr formulation

---

[6] Equation (5.30) is the correct solution of Equation (5.29) provided by Prácser and Szarka (1999). It should replace the incorrect solution by Bahr (1991).

or the Groom and Bailey formulation is the requirement that the distortion tensor should be real and frequency independent. As for the Bahr technique, electric, rather than magnetic distortion of the impedance tensor is considered to be of greatest significance in the Groom and Bailey decomposition model.

In Groom and Bailey's decomposition technique, separation of the *'localised'* effects of 3-D current channelling from the *'regional'* 2-D inductive behaviour is achieved by factorising the impedance tensor problem in terms of a rotation matrix, $\underline{\underline{\beta}}_\alpha$, and a distortion tensor, $\underline{\underline{C}}$, which is itself the product of three tensor suboperators (twist, $\underline{\underline{T}}$, shear, $\underline{\underline{S}}$, local anisotropy, $\underline{\underline{A}}$,[7] and a scalar, g,:

$$\underline{\underline{Z}} = \underline{\underline{\beta}}_\alpha\,\underline{\underline{C}}\,\underline{\underline{Z}}_{2D}\underline{\underline{\beta}}_\alpha^T \quad \text{where} \quad \underline{\underline{C}} = g\underline{\underline{T}}\,\underline{\underline{S}}\,\underline{\underline{A}}. \tag{5.33}$$

Groom and Bailey illustrate the need for the four independent parameters in the distortion tensor factorisation by considering a scenario in which MT data are collected at the centre of an elliptical swamp, which presents a highly conductive surface region, encompassed by an insulating substratum in a moderately conductive region (Figure 5.5). The presence of the swamp causes the telluric currents to be twisted to its local strike. This clockwise rotation of the telluric vectors is contained in the twist tensor, $\underline{\underline{T}}$. The swamp also causes splitting of the principal impedances by generating different distortion-related stretching factors. This anisotropic effect is contained in the anisotropy tensor, $\underline{\underline{A}}$. Meanwhile, the shear tensor, $\underline{\underline{S}}$, stretches and deflects the principal axes so that they no longer lie orthogonal to each other. Whereas $\underline{\underline{A}}$ does not change the directions of telluric vectors lying along either of the principal axes, $\underline{\underline{S}}$ generates the maximum angular deflection along these principal axes. In the case of strong current channelling, the direction of the variational electric field does not depend on the direction of the magnetic field at all. Distortions manifest as non-orthogonal telluric vectors may also arise owing to inaccurate alignment of telluric dipoles during site setup.

Equation (5.33) presents an ill-posed problem as it contains nine distinct unknown parameters – the regional azimuth, the four components of the distortion tensor and the two complex regional impedances – whereas the complex elements of the measured impedance tensor provide only eight known parameters. Therefore, the

---

[7] This is a distortion effect and does not have the same physical meaning as 'electrical anisotropy' as proposed, for example, by Kellet *et al.* (1992). Electrical anisotropy is discussed with reference to case studies in Chapter 9.

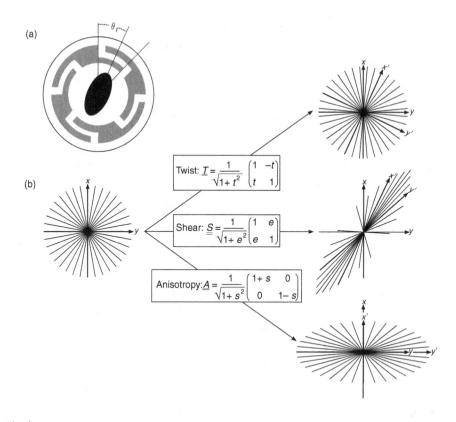

Figure 5.5 (a) A contrived scenario in which MT data are collected at the centre of a conductive swamp (black) that is encompassed by a moderately conductive region (grey), and an insulator (white). θ_t denotes the local strike of the swamp, which 'twists' the telluric currents. The anomalous environment also imposes shear and anisotropy effects on the data. (b) Distortion of a set of unit vectors by twist $\underline{\underline{T}}$, shear, $\underline{\underline{S}}$, and 'anisotropy', $\underline{\underline{A}}$, operators, which are parameterised in terms of the real values t, e and s, respectively. (Redrawn from Groom and Bailey, 1989.)

decomposition model as posed in Equation (5.33) does not have a unique solution. However, a unique solution does exist if the linear scaling factor, g, and anisotropy, $\underline{\underline{A}}$, are absorbed into an equivalent ideal 2-D impedance tensor that only differs from $Z_{\text{2-D}}$ in that it is scaled by real, frequency-independent factors:

$$\underline{\underline{Z}}'_{\text{2-D}} = g\underline{\underline{A}}\,\underline{\underline{Z}}_{\text{2-D}}. \tag{5.34}$$

This transformation leaves the shapes of the apparent resistivity and impedance phase curves unchanged, but the apparent resistivity curves will be shifted by unknown scaling factors. Once again (cf. Larsen's model), this is equivalent to stating that there is no way of determining static shift factors mathematically from a measured impedance.

Although a unique solution of the decomposition model exists theoretically, in practice measured data, which contain noise and departures from the model, will never yield a perfect fit to the decomposition model in any co-ordinate system. Therefore a least-square (Section 4.2) fitting procedure is employed. In Groom–Bailey decomposition, a misfit parameter between the measured data, $\underline{\underline{Z}}_{ij}$, data errors, σ_{ij}, and data modelled according to the 2-D hypothesis, $\underline{\underline{Z}}_{ij,m}$ is suggested as:

$$\chi^2 = \frac{1}{4}\frac{\sum\limits_{j=1}^{2}\sum\limits_{i=1}^{2}|\hat{Z}_{ij,m} - Z_{ij}|^2}{\sum\limits_{i=1}^{2}\sum\limits_{j=1}^{2}|\sigma_{ij}|^2}. \tag{5.35}$$

3-D induction effects are not considered directly in the decomposition parameterisation, but their presence can be detected by investigating distortion parameters and misfit parameters. In the presence of 3-D induction, the computed distortion parameters will exhibit frequency dependence, and misfit parameters can be expected to be large. However, note that χ^2 is normalised by the errors in the data. So, a changing error structure in different period ranges may influence the misfit.

Decomposition models (Groom and Bailey, 1989; Bahr, 1991; Chave and Smith, 1994; and Prácser and Szarka, 1999) have been applied to many datasets. Figure 5.6 summarises the seven classes of impedance tensor that have been proposed. The parameters (such as regional strike) for the more-complicated models can be unstable if a measured impedance can already be described by a simpler general model. For example, Equation (5.22) will not yield stable results if an impedance tensor can be described by Equation (5.19) (Berdichevsky, 1999). In cases of strong current channelling or 3-D induction, the electromagnetic strike obtained using mathematical decomposition is also unstable.

The decomposition model described in Equation (5.22) was adopted by Kellet *et al.* (1992), who suggested that the existence of an electromagnetic strike and the impedance phase difference associated with it can be a consequence of electrical anisotropy in a particular depth range. Kellet *et al.* (1992) proposed a physical model that comprises a superposition of a surface scatterer and a 1-D electrically anisotropic regional structure (Figure 5.7). Although developed for a limited target area, Kellet *et al.*'s model may explain why decomposition models can apparently be used to describe so many different datasets – given that a 1-D anisotropic structure is mathematically equivalent to a 2-D structure. An example of the impedance phase split created by an anisotropic structure is presented in Figure 5.8. The concept of lower crustal electrical anisotropy in connection with the impedance tensor decomposition model was used by Jones *et al.* (1993) and Eisel and Bahr (1993) to explain the data shown in Figure 5.8, with resistivity ratios of up to 60 between maximum and minimum directions of resistivity in

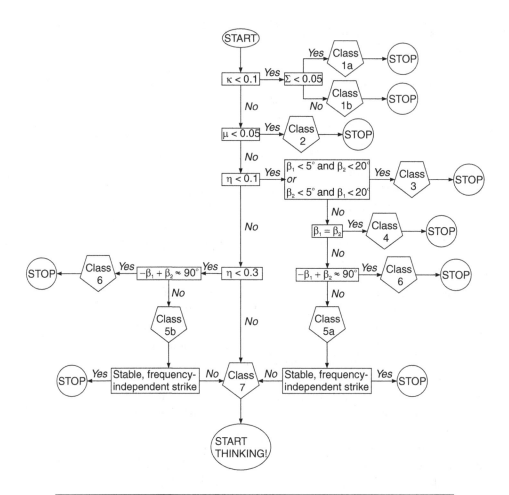

Class 1a : Simple 1-D model (Cagniard, 1953), Equation (5.9)
Class 1b : Simple 2-D model (Swift, 1967), Equation (5.11)
Class 2 : Regional 1-D model with galvanic distortion (Larsen, 1977), Equation (5.19)
Class 3 : 2-D model with static shift (Bahr, 1991), Equation (5.22)
Class 4 : Regional 2-D model in 'twisted' co-ordinates (Bahr, 1991), Equation (5.22)
Class 5a : 2-D superimposition model (Bahr, 1988), Equation (5.22)
Class 5b : 2-D Extended superimposition (δ-) model (Bahr, 1991), Equation (5.28)
Class 6 : Regional 2-D model with strong local current channelling (Bahr, 1991)
Class 7 : Regional 3-D induction model

Figure 5.6 Flowchart summarising characterisation and model parameterisation for different classes of the impedance tensor.

the horizontal plane. Bahr *et al.* (2000) suggested a general 'anisotropy test' based on the impedance phases present in electromagnetic array data, and showed that this test could distinguish between the model of crustal anisotropy and the model of an isolated conductivity anomaly.

In the course of applying decomposition techniques, care should be taken to consider the appropriateness of the technique

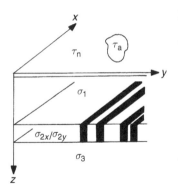

Figure 5.7 Kellet *et al.*'s (1992) superimposition model in which a small-scale, near-surface conductance anomaly, τ_a, scatters electric fields, distorting the measured impedance away from that of the regional 1-D model containing an anisotropic layer, represented by a series of vertical dykes with alternating resistivities.

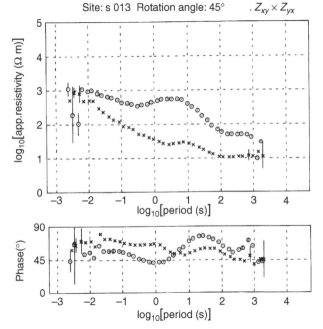

Figure 5.8 Apparent resistivities and impedance phases from an MT site in British Columbia (BC87 dataset). The data have been interpreted in terms of a superimposition model similar to that shown in Figure 5.7, with the impedance phase split being ascribed to an anisotropic structure in the lower continental crust (Jones *et al.*, 1993; Eisel and Bahr, 1993).

for the dataset being analysed. If only one impedance phase is present in the measured impedance tensor, then the inductive structure is 1-D and there is no justification for the decomposition hypothesis. If the data exhibit minimal impedance phase splitting and large errors, then a stable strike is unlikely to be retrieved. The errors of the phase-sensitive skew and phase-sensitive strike are larger than the Swift skew and Swift strike, because the errors propagate through more computational steps. The impedance phase splitting is incorporated in the error of the phase-sensitive skew via these relationships:

$$\Delta\eta^2 = \left(\frac{\Delta C}{2\sqrt{CD_2}}\right)^2 + \left(\frac{\Delta D_2 \sqrt{C}}{D_2^2}\right)^2 = \left(\frac{(\Delta C)^2}{4D_2^4\eta^2}\right) + \left(\frac{\eta^2(\Delta D_2)^2}{D_2^2}\right), \quad (5.36)$$

where

$$C = [D_1, S_2] - [S_1, D_2]$$

and

$$(\Delta C)^2 = (\Delta D_1)^2|S_2|^2 + (\Delta S_2)^2|D_1|^2 + (\Delta S_1)^2|D_2|^2 + (\Delta D_2)^2|S_1|^2.$$

Here, Δ denotes the error of the function that it precedes.

The following confidence interval of the phase-sensitive strike ($d\alpha$) incorporates both errors in the measured data and the degree of phase splitting (Simpson, 2001b):

$$\Delta\alpha = \frac{1}{2}\frac{1}{1+(\tan 2\alpha)^2}\Delta(\tan 2\alpha), \quad (5.37)$$

where $[\Delta(\tan 2\alpha) =$

$$\sqrt{\left(\frac{\Delta([S_1,S_2]-[D_1,D_2])}{[S_1,D_1]+[S_2,D_2]}\right)^2 + \left(\frac{\Delta\{([S_1,D_1]+[S_2,D_2])([S_1,S_2]-[D_1,D_2])\}}{([S_1,D_1]+[S_2,D_2])^2}\right)^2},$$

$$\{\Delta([S_1, S_2] - [D_1, D_2])\}^2 = \Delta S_1^2 S_2^2 + S_1^2 \Delta S_2^2 + \Delta D_1^2 D_2^2 + D_1^2 \Delta D_2^2$$

and

$$\{\Delta([S_1, D_1] + [S_2, D_2])\}^2 = \Delta S_1^2 D_1^2 + S_1^2 \Delta D_1^2 + \Delta S_2^2 D_2^2 + S_2^2 \Delta D_2^2.$$

An example of error propagation in synthetic data with a constant error floor of 2% is presented in Figure 5.9. The key points to note from this figure are:

(i) errors in the phase-sensitive skew and phase-sensitive strike are clearly larger than the errors in the Swift skew and Swift strike (Equations (5.17) and (5.18)), respectively; and

(ii) the errors in the phase-sensitive skew and phase-sensitive strike are largest where the impedance phase splitting is small and smaller where the impedance phase splitting increases.

An example of error propagation in measured data is presented in Figure 5.10. In this case, the errors in the data increase at periods longer than 1000 s, and we see a corresponding increase in the errors of the phase-sensitive skew and phase-sensitive strike. In the ~100 s to ~1000 s period range, small data errors combine with large impedance phase differences to yield well-constrained decomposition

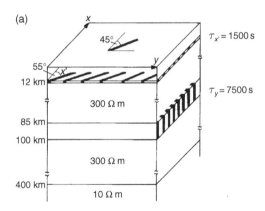

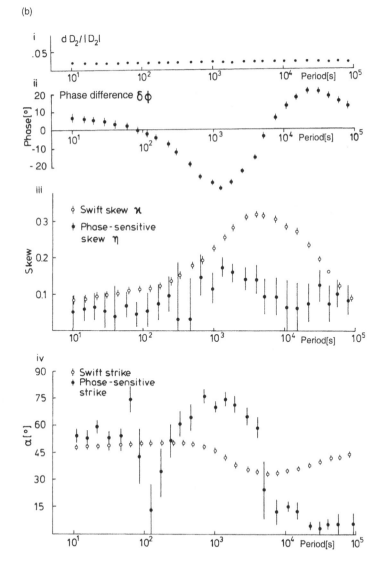

Figure 5.9 (a) 1-D layered model incorporating two anisotropic layers (with their tops at 12 km and 85 km, respectively) with different strikes. (b) Propagation of (i) 2% error imposed on synthetic impedance data derived from model shown in (a), (which gives rise to the phase differences, $\delta\phi$, between the principal polarisations of the impedance tensor shown in (ii)) into (iii) Swift skew, κ, and phase-sensitive skew, η, and (iv) the Swift strike and phase-sensitive strike.

Figure 5.10 Propagation of (i) errors in measured impedances having the phase differences between the principal polarisations of the impedance tensor shown in (ii) into (iii) *Swift skew, κ, and phase-sensitive skew, η,* and (iv) the Swift strike and phase-sensitive strike. The combination of small errors and large impedance phase differences, $\delta\phi$, in the 100–1000 s period range results in well-constrained estimates for the skew and strike angles in this period range.

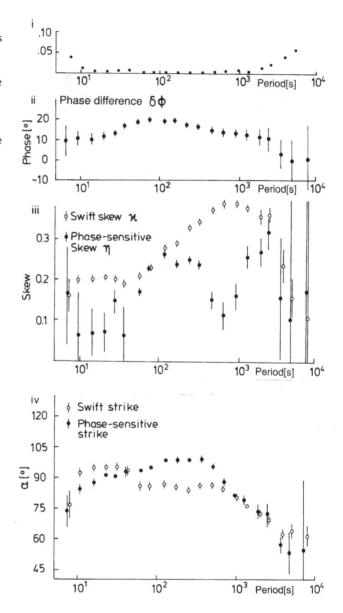

parameters, whereas at periods shorter than $\sim$100 s, the impedance phase difference is smaller and the errors become larger.

5.4 Problems with decoupling E- and B-polarisations: farewell to the 2-D world

Figure 5.11 shows how the inductive responses of conductivity anomalies having finite strike extents compare with those from

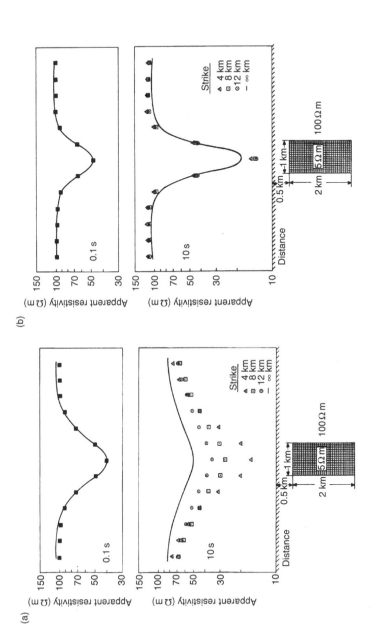

Figure 5.11 Comparison, at 0.1 s and 10 s, of (a) **E**-polarisation and (b) **B**-polarisation responses arising from a 3-D 5 Ω m conductivity anomaly embedded in a 100 Ω m half-space and having finite strike extents of either 4 km, 8 km or 12 km with those owing to a 2-D (i.e., extending to infinity in the strike direction) conductivity anomaly exhibiting the same 5:100 Ω m resistivity contrast. At 0.1 s all of the 3-D responses (black squares) are close to those of the 2-D model, because galvanic effects are insignificant compared to inductive effects, and because inductive fields from the ends of the prism are severely attenuated. (Redrawn from Ting and Hohmann, 1981.)

2-D (extending to infinity in the strike direction) conductivity anomalies. The effect of replacing a 2-D conductivity anomaly with one having finite strike is more significant for the **E**-polarisation than for the **B**-polarisation (Ting and Hohmann, 1981). For example, at 10 s, a significant dip in **E**-polarisation apparent resistivity (Figure 5.11(a), lower panel), occurs as a 5 Ω m conductivity anomaly extending only 12 km along strike is crossed compared to the **E**-polarisation apparent resistivity response from a 5 Ω m conductivity anomaly of infinite strike extent. As the strike extent of the conductivity anomaly is reduced, the deviation of the 3-D apparent resistivity from the 2-D **E**-polarisation apparent resistivity increases. The 3-D anomalies imaged by the **E**-polarisation appear more conductive than the 2-D anomaly, although all conductors in the model have the same resistivity. The **B**-polarisation apparent resistivity (Figure 5.11(b), lower panel), on the other hand, is less dramatically effected. This has led to a widespread notion, often quoted in electromagnetic literature, that because the **B**-polarisation is 'robust' in this sense, 2-D inversion of just the **B**-polarisation data, whilst ignoring the **E**-polarisation data is an acceptable practice. What seemed an acceptable compromise for avoiding some of the effects of three-dimensionality in the 1980s, prior to viable 3-D modelling schemes, is now known to be misguided. Data acquisition costs a lot of time, money and effort, and there's no excuse for discarding upwards of 75% of it! Departures from two-dimensionality often provide the most interesting information about the Earth, and even a 1-D anisotropic structure cannot be diagnosed if only **B**-polarisation data are considered!

For the case of the comparisons shown in Figure 5.11, the conductivity anomaly has an unambiguous strike direction, whereas for the case of measured data, we might also be confronted with a scenario in which more than one strike exists in different depth ranges. If a stable strike does not exist, it will be impossible to define a single co-ordinate frame in which the entire period range of the data can be satisfactorily decoupled into **E**- and **B**-polarisations. If the data are rotated to an average strike and modelled two-dimensionally then the resulting 2-D model may contain artefacts at depth that arise from structure that is laterally displaced from the profile. On the other hand, the period dependence of the electromagnetic strike can be an important diagnostic tool for decoupling regional structures. This is demonstrated by a case study from NE England (Figure 5.12). Figure 5.12(a) shows the electromagnetic strike (determined using Groom–Bailey decomposition (Groom and Bailey, 1989) plotted as a function of period. At short

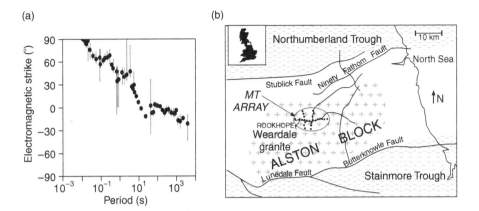

periods – from 10^{-2} s to 10^{-1} s – the strike is $\sim70°$, whilst at long periods – from 100 s to 1000 s – an approximately 0° strike can be inferred, and from 10^{-1} s to 100 s the strike exhibits a large frequency dependence. Considering the regional environment (Figure 5.12 (b)), how would you explain this period dependence of the electromagnetic strike? What do you imagine the outcome would be if an average strike were taken, and a 2-D inversion were performed on these data? We shall return to this case study in more detail in Section 7.3.

Figure 5.12 (a) Period-dependent electromagnetic strike (determined using Groom–Bailey decomposition (Groom and Bailey, 1989) for a site collected within (b) an MT array located in NE England. (Redrawn from Simpson and Warner, 1998.)

5.5 Decomposition as an hypothesis test: avoiding the black-box syndrome

In 1899, the geologist Chamberlain wrote:

> The fascinating impressiveness of rigorous mathematical analyses, with its atmosphere of precision and elegance should not blind us to the defects of the premises that condition the whole process. There is perhaps no beguilement more insidious and dangerous than an elaborate and elegant mathematical process built upon unfortified premises.

Chamberlain (1899) wrote this as a rebuttal to Lord Kelvin's estimate for the age of the Earth (which he based on the concept that the Earth cooled by conduction from a molten iron sphere) of less than 100 million years. However, this wonderfully eloquent statement is also worth bearing in mind when interpreting MT data with the aid of certain techniques which have achieved the sometimes dubious status of 'standard'. A label that in a scientific sense is somehow contradictory, and possibly a threat to progress.

Decomposition methods were originally advanced as **hypothesis tests** for the idea that the regional, deep, geoelectric structure could be treated as 2-D, after allowing for purely galvanic distortion owing to near-surface heterogeneities. In practice, decomposition methods have become widely employed more as black box methods for establishing an average electromagnetic strike direction along a profile of MT stations, because the ultimate aim of many users is 2-D inversion of the data. This is lamentable, because acknowledging the frequency dependence of the electromagnetic strike can potentially provide us with an important complementary parameter that can be related to depth or lateral displacement. When applied as an hypothesis test, impedance tensor decomposition can facilitate decoupling of structures lying laterally closer, or shallower, on the one hand, and laterally further away, or deeper, on the other (e.g., Section 5.6). The potential for MT data to provide well-constrained information concerning structural strike is a forte of the MT method, and a major advantage over seismic methods. Some applications of the electromagnetic strike as a diagnostic tool include mapping faults, and investigating mantle flow (see Chapter 9).

Again, we emphasise that if the decomposition hypothesis holds, then we should obtain stable frequency-independent decomposition angles. If your data do not yield frequency-independent angles, this doesn't mean that decomposition hasn't worked. Decomposition has worked, but the results that decomposition have yielded are indicating that the hypothesis being tested – that the regional structure is 2-D – is incorrect. Decomposition is a tool. It's a useful tool if applied correctly – as an **hypothesis test**, and not as a black box – but it won't make your data 2-D if it isn't. So, when you consider how to interpret the results of decomposition, spare a thought for a nineteenth-century geologist.

5.6 How many strike directions are there?

Since 'regional' and 'local' scales can be different for different electromagnetic sounding periods, it is clear that whilst the decomposition hypothesis may fail for some period ranges of a dataset, it may yet be satisfied for more limited period ranges. Dynamic processes related to plate tectonics continually modify the Earth's structure. Whilst some plates break apart, others collide and merge; plates alter their directions of motions; stress fields rotate. In this balance of destruction and renewal, remnants of past tectonic events are often preserved, giving rise to tectonic environments in which structural lineaments trend in more than one direction. One such environment exists in southern Kenya, where a major NW–SE trending suture zone between the Archaean Nyanza craton and the Mozambique Mobile Belt coexists with the N–S trending Rift Valley.

In 1995, MT data were acquired along an E–W profile crossing the N–S trending Kenya rift (Figure 5.13 (a)). Assuming a 2-D electromagnetic structure controlled by the rifting process, a profile of sites perpendicular to the rift seemed to be an appropriate experimental configuration. Contrary to expectation, the electromagnetic strike was found to be both period- and site-dependent. Close to the western boundary of the rift marked by a 1500-m escarpment, a N–S strike was recovered, but away from the escarpment, on the western flank of the rift, an approximately NW–SE strike was detected, followed by a second N–S trending boundary (delineated by the Oloololo escarpment). Close to the eastern margin of the rift, both the N–S strike of the rift and an approximately NW–SE strike influence the data (Figure 5.13 (b)). Both N–S and NW–SE striking structures are also reflected strongly in the induction arrows (Figure 5.13 (c)).

The electromagnetic strike hinted at structural trends that were not detectable from seismic refraction or gravity surveys previously

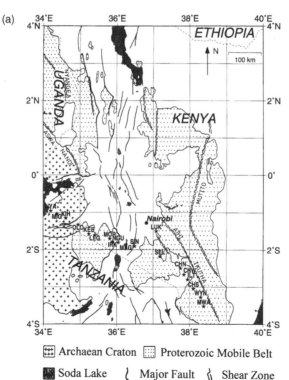

(a)

Figure 5.13 (a) Geological map of East Africa showing locations of MT sites (indicated by stars) collected in 1995 as part of the Kenya Rift International Seismic Project (KRISP). (b) Pseudosection showing variations in decomposition angle (Groom and Bailey, 1989) as a function of period and distance along a profile crossing the southernmost part of the Kenya rift. (c) Real (*solid lines*) and imaginary (*dotted lines*) induction arrows at MT sites shown in (a) for discrete periods between 64 s and 4096 s (Simpson, 2000).

⊞ Archaean Craton   ▦ Proterozoic Mobile Belt

◼ Soda Lake   ⸨ Major Fault   ⸩ Shear Zone

(b)

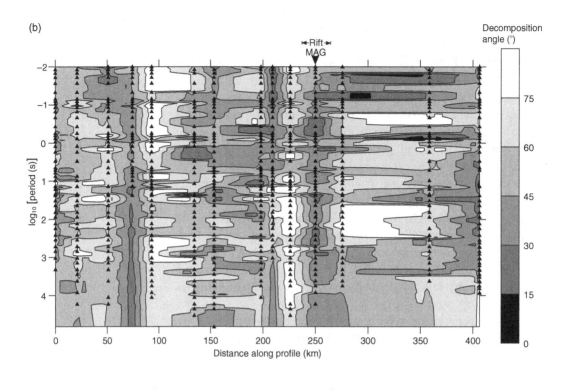

(c)

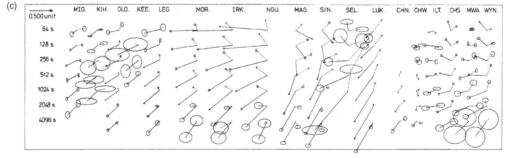

Figure 5.13 (cont.).

carried out along the same profiles, but posed a problem (analogous to side-swipe in seismic techniques) with regard to 2-D modelling: because MT provides a volume sounding, a 2-D model can contain artefacts at depth that are actually due to structure that is laterally displaced from the profile, unless the data are rotated into an appropriate co-ordinate frame that decouples **E**- and **B**-polarisations. In this case study, the profile runs sub-parallel to one of two dominant regional strikes, making adequate decoupling and subsequent 2-D modelling of the data untenable. Instead, 3-D forward modelling was used to demonstrate that the NW–SE electromagnetic strike probably relates to shear zones delineating the suture

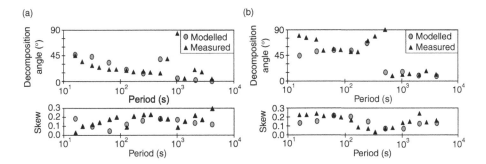

zone between the Mozambique Mobile Belt and Archaean Nyanza craton that collided during the Proterozoic (Simpson, 2000). A model consisting of the near-surface overburden of the N–S striking rift, and more deeply penetrating NW–SE conductive lineaments, could explain not only the induction arrows, apparent resistivities and impedance phases, but also the period and site dependence of phase-sensitive strike and skew (Figure 5.14). Site MAG (Magadi), within the rift, exhibits an anomalous strike that is neither N–S nor NW–SE. Comparison of the measured and modelled decomposition angles (Figure 5.14) reveals the anomalous strike to be a virtual one, arising from coupling between the N–S and NW–SE strikes. At a neighbouring site (SIN) closer to the eastern flank of the rift, N–S and NW–SE strikes are retrieved in different period ranges.

The Kenya case study demonstrates that appropriate application of the impedance tensor decomposition hypothesis can facilitate decoupling of structures lying laterally closer, or shallower, on the one hand, and laterally further away, or deeper, on the other. Therefore, we shouldn't be perturbed when we see a period- or location-dependent strike, or try to average out the differences that may be apparent. MT data can potentially constrain directionality and its source better than seismological data – something that should be viewed as an asset, rather than as an inconvenience of the MT method. However, we have to exercise caution, because, in certain circumstances, the electromagnetic strike retrieved by impedance tensor decomposition may be a virtual strike only, as in the case of site MAG.

Figure 5.14 Measured and modelled decomposition angles (phase-sensitive 'strikes'[8]) and phase-sensitive skews at (a) an MT site (MAG) close to the axis of the Kenya rift (Figure 5.13); and (b) an MT site (SIN) close to the eastern boundary of the rift. (Redrawn from Simpson, 2000.)

---

[8] Inverted commas around 'strike' are in recognition of the fact that, strictly speaking, a strike belongs to a 2-D structure, whereas the period dependence of the decomposition angle belies the idea of two-dimensionality.

5.7 The concepts of anisotropy, skew bumps, and strike decoupling

In 1977, Reddy *et al.* observed that, 'Stations with skew values greater than 0.2 are significantly influenced by the three-dimensionality of the geoelectric structure'. Unfortunately, this statement has subsequently been misconstrued to falsely infer that if skew values are less than 0.2, then the conductivity structure is essentially 2-D. Figure 5.15 shows skew values in the vicinity of a $10\,\Omega\,m$ box embedded in a $100\,\Omega\,m$ host. Most of the contours highlighted in Figure 5.15 have skew values less than 0.2 and, along the axis of symmetry, skew values are zero (Reddy *et al.*, 1977).

In 1991, Bahr discussed what he classified as Class 7 distortion – a regional 3-D anomaly – writing 'This class includes those cases where $\eta > 0.3$, and therefore even the regional conductivity distribution is not 2-D'. What is the most important word in this sentence? The key word here is 'includes'. In other words, the condition that $\eta < 0.3$ is a necessary but not sufficient indication of a regionally 2-D structure.

In the MT case study from southern Kenya (Section 5.6), we demonstrated that the data are influenced by three-dimensionality. Note that despite the existence of differently striking structures, which at station MAG are inductively coupled, the phase-sensitive skew, η, is less than 0.3 (Figure 5.14). Thus, the condition that $\eta < 0.3$ is clearly an unreliable indicator of two-dimensionality.

A further example of the unreliability of skew values for diagnosing the dimensionality of geoelectric structure is shown in

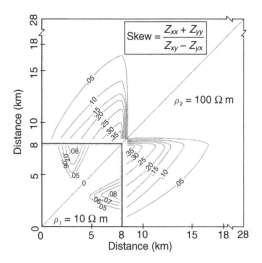

Figure 5.15 Contours showing swift skew values in the vicinity of a 10 Ω m cube embedded in a 100 Ω m half-space. Skew is zero along the axis of symmetry indicated by the diagonal line. (Redrawn from Reddy *et al.*, 1977.)

$$\text{Skew} = \frac{Z_{xx} + Z_{yy}}{Z_{xy} - Z_{yx}}$$

$\rho_2 = 100\ \Omega\ m$

$\rho_1 = 10\ \Omega\ m$

Distance (km)

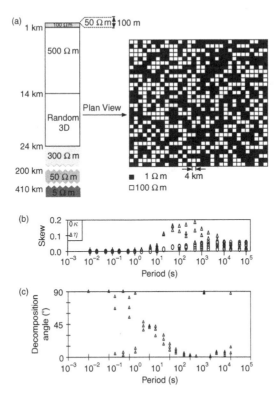

Figure 5.16 (a) Section of a 3-D layer that extends throughout the depth range 14 to 24 km in an otherwise 1-D layered model; (b) the Swift skew, κ, and Bahr's phase-sensitive skew, η, are less than 0.2, at three sites stationed above the centre of the 3-D layer, although the three-dimensionality is sufficient to generate (c) a period-dependent decomposition angle.

Figure 5.16. Random three-dimensionality has been generated in the depth range 14 km to 24 km with a simple condition that cells 4 km × 4 km square have conductivities of $1\,\Omega$m or $100\,\Omega$m, with equal 50:50 probability. Both the Swift skew (κ), and Bahr's phase-sensitive skew (η) are shown to be less than 0.2 at three example sites stationed over the centre of the grid, although the three-dimensionality is sufficiently significant to generate a period-dependent electromagnetic strike.

Based on a dataset exhibiting negligible induction arrows, and uniform impedance phase splitting along a profile of sites, Kellet *et al.* (1992) interpreted the deep crust below the Abitibi Belt in Canada to be anisotropic. The impedance phase splitting was attributed to anisotropy at depth rather than to lateral conductivity anomalies, because lateral conductivity gradients generate vertical magnetic fields, and because the amount of impedance phase splitting should decrease with increasing distance from any lateral structure. Kellet *et al.* (1992) represented the anisotropy in their model with a series of alternating vertical dykes with two different resistivity values (see Figure 5.7). However, for MT purposes, the alternating dyke model shown in Figure 5.7 is equivalent to the

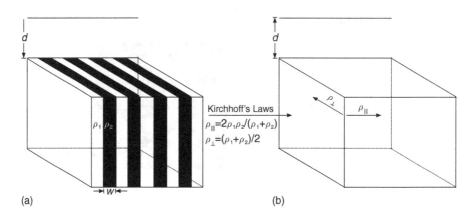

(a)   (b)

Figure 5.17 (a) Macroscopic anisotropy, and (b) intrinsic anisotropy. For $d > w$, the models shown in (a) and (b) yield MT impedances that are equivalent (i.e., macroscopic anisotropy is indistinguishable from intrinsic anisotropy at depth).

intrinsic anisotropy model shown in Figure 5.17, because MT transfer functions are unable to resolve conductivity variations that have horizontal dimensions less than (or even up to several times) the depth at which they are imaged. In other words, MT transfer functions cannot distinguish between macroscopic and microscopic electrical anisotropy. The inherent lack of lateral resolution for structures at depth is discussed in Section 4.5, and illustrated in Figure 4.8.

A 1-D anisotropic layer should have zero skew. However, in measured data we generally observe a non-zero skew. So, if we want to refine our models then we might consider how skew is generated. One possibility involves an anisotropic signature with a direction that meanders. For the example shown in Figure 5.18, a bump in the phase-sensitive skew is generated in the period range ~ 10–10^4 s. Of course, this isn't the only way to generate a 'skew bump' – we might also consider multi-layer anisotropy with different layers having different strikes. In this case, in addition to generating a skew bump, we should observe a period-dependent strike, with the shorter-period strike

Figure 5.18 (a) Model containing an anisotropic layer composed of meandering dykes, and (b) skew generated by the anisotropic layer.

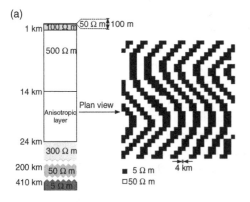

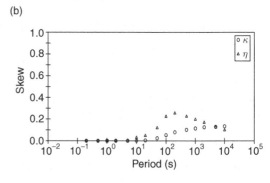

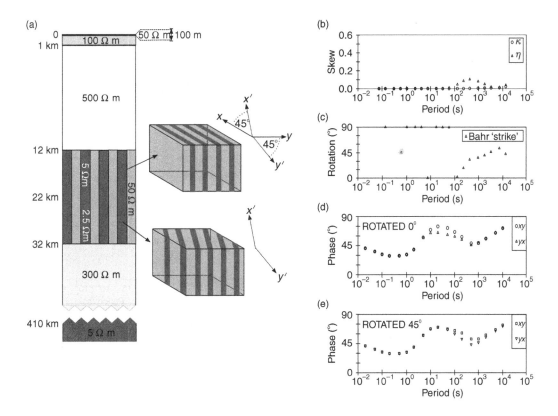

corresponding to the shallower layer, the longer-period strike corresponding to the deeper layer, and a transition region in between, where the recovered strike is unstable (Figure 5.19).

Having introduced these slightly more complex models of anisotropy, we should, however, take some time to reflect on what we mean by 'anisotropy'. Although a model containing anisotropy leads to anisotropic data, in which the phases of the two principal polarisations are split, anisotropic data doesn't necessarily imply a model containing anisotropy!

If the series of dykes shown in Figure 5.7 is replaced by a dipping interface with resistivities of $50\,\Omega$m on one side and $5\,\Omega$m on the other, then two extensive regions for which the impedance phase splitting is approximately equivalent arise on either side of the boundary (Figure 5.20). Only across a narrow region close to the boundary can the convergence, cross-over and divergence of the phases of the two polarisations be observed. Depending on how MT sites are distributed, the impedance phase splitting generated by the dipping boundary might therefore be falsely ascribed to two different anistropic regimes.

Figure 5.19 (a) Model containing two anisotropic layers with different strikes; (b) skew generated by (a); (c) period-dependent phase-sensitive strike; (d) and (e) impedance phases synthesised at the surface of the model, and rotated through (d) 0° (strike of upper anisotropic layer) and (e) 45° (strike of lower anisotropic layer).

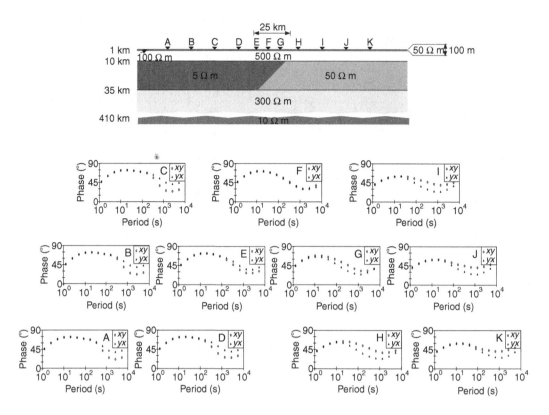

Figure 5.20 Impedance phase splitting at sites lying along a profile crossing a dipping contact. Notice that sites A–D have similar phases, as do sites H–K, and that (unlike for the case of a vertical contact), the phase splitting diminishes proximate to the contact. These impedance phases might be falsely interpreted in terms of two adjunct anisotropic layers with orthogonally striking directions of high conductivity.

Having introduced anisotropy into our models, we should also spare some thought for whether anisotropy is useful in terms of advancing physically meaningful interpretation of data, or whether it's just a convenient way of modelling data that images different conductances in different polarisations. Does anisotropy help to constrain possible conduction mechanisms, or does it just help us to reduce the misfit involved in jointly fitting MT data within the framework of a 2-D model, whilst ignoring 3-D lateral effects? Adding anisotropy to a model is certainly the easiest way to obtain a fit to data exhibiting different conductances in different directions, but the introduction of anisotropy needs to be justified. Such justification is most easily demonstrated when array data, rather than data from a profile of sites are available, because the possibility that impedance phase splitting is caused by conductivity contrasts that are laterally offset from an individual profile can be ruled out more easily where array data are available. We will return to the concept of anisotropy in conductivity models in Chapters 6–9.

5.8 Static shift, the long and the short of it: 'correcting' or modelling?

Static shift causes a frequency-independent offset in apparent resistivity curves so that they plot parallel to their true level, but are scaled by real factors. The scaling factor(s) or static shift factor(s), s^9, cannot be determined directly from MT data recorded at a single site. A parallel shift between two polarisations of the apparent resistivity curves is a reliable indicator that static shift is present in the data. However, a lack of shift between two apparent resistivity curves does not necessarily guarantee an absence of static shift, since both curves might be shifted by the same amount. The correct level of the apparent resistivity curves may lie above, below or between their measured levels. If MT data are interpreted via 1-D modelling without correcting for static shift, the depth to a conductive body will be shifted by the square root of the factor by which the apparent resistivities are shifted ($\sqrt{s}$), and the modelled resistivity will be shifted by s. 2-D and 3-D models may contain extraneous structure if static shifts are ignored. Therefore, additional data or assumptions are often required.

Static shift corrections may be classified into three broad methods.

(i) Short-period corrections relying on active near-surface measurements (e.g., TEM, DC).
(ii) Averaging (statistical) techniques.
(iii) Long-period corrections relying on assumed deep structure (e.g., a resistivity drop at the mid-mantle *transition zones*) or long-period magnetic transfer functions.

(i) Active EM techniques such as *transient electromagnetic sounding* (*TEM*) do not have the static shift problems that afflict MT sounding. Therefore, in some environments TEM data can be used in conjunction with MT data from the same site in order to correct for static shifts in MT data. Apparent resistivity curves derived from MT data can be scaled so that they agree with TEM curves, or jointly inverted with TEM data (Sternberg *et al.*, 1988). Near-surface resistivities can also be constrained using DC data (e.g., Simpson and Warner, 1998). These techniques are only effective for correcting static shift arising from near-surface

---

[9] Here and in the following discussion we use '*s*' to denote the factor by which apparent resistivities are scaled. Some authors refer to the static shift of the impedance, which in our formalism is $\sqrt{s}$.

heterogeneities; in multi-dimensional environments shifting both polarisations of the MT apparent resistivity curves to an average level determined from TEM soundings or DC models is not always appropriate. For example, if the impedance phases are split at the high frequencies that are expected to have similar penetration depths to TEM or DC data, then there is no justification for supposing that the apparent resistivity curves for the two polarisations should lie at the same level. Active EM techniques are discussed further in Section 10.5.

(ii) Averaging (statistical) techniques tend to give relative, rather than absolute values of static shift. Relative static shift corrections help to preserve the correct form of an anomaly in a multi-dimensional model, whilst properties such as conductance and depth may be inaccurate. Inaccuracies in conductance and depth estimates based on averaging techniques are expected to decrease as the site density and the size of the region mapped increase.

Electromagnetic array profiling (EMAP) is an adaptation of the MT technique that involves deploying electric dipoles end-to-end along a continuous profile. In this way, lateral variations in the electric field are sampled continuously and spatial low-pass filtering can be used to suppress static shift effects (Torres-Verdin and Bostick, 1992). This method of static shift correction relies on the assumption that the frequency-independent secondary electric field, E_s, that arises owing to boundary charges along the profile sums to zero as the length of the profile tends to infinity, i.e.,

$$\int_0^\infty E_s = 0. \tag{5.38}$$

In practice, the finite length of the profile may lead to biases, particularly if the profile terminates on an anomaly.

In some statistical techniques, as for example those sometimes employed in the *Occam 2-D inversion* algorithms (de Groot-Hedlin, 1991), it is assumed that the mean average static shift of a dataset should be unity (i.e., a static shift factor at a particular site is cancelled by the static shift factor, s, calculated at (an) other site(s)):

$$\sum_i^{N_{SITES}} \log_{10}(s) = 0. \tag{5.39}$$

If there are more small-scale conductive heterogeneities than resistive ones (or vice versa) there will be a systematic bias in the static shifts estimated assuming the zero-mean random log distribution expressed in Equation (5.39). This is particularly so for datasets with only a small number of sites. Ogawa and Uchida (1996) suggest, instead, that the static shifts in a dataset should form a Gaussian distribution. This is consistent with, but does not require a mean static shift of unity. For datasets with a small number of sites, the assumption that the static shift factors form a Gaussian distribution is probably more appropriate than the assumption that the static shift factors should sum to unity. However, there is no rigorous basis for either of these assumptions.

Other averaging techniques rely on the idea of shifting apparent resistivity curves so that the longest periods ($>10^4$ s) correspond with 'global' or regional apparent resistivity values that are computed from geomagnetic observatory data. Geomagnetic transfer functions such as Schmucker's *C-response* (see Section 10.2) are free of static shift. This technique may not be appropriate where a high level of heterogeneity is present in the mantle, or where high-conductance anomalies in the crust limit the penetration depth that is achieved.

(iii) An *equivalence relationship* can be shown to exist (Schmucker, 1987) between the MT impedance tensor, $\underline{Z}$, and Schmucker's C-response (Appendix 2):

$$\underline{Z} = i\omega\underline{C}. \qquad (5.40)$$

Schmucker's C-response can be determined from the magnetic fields alone, thereby providing an inductive scale length that is independent of the distorted electric field. Therefore the equivalence relationship given in Equation (5.40) can be used to correct for static shift in long-period MT data. Magnetic transfer functions can be derived from the same *time series* as the MT transfer functions. Magnetic transfer functions can, for example, be derived from *solar quiet (Sq) variations*, which are daily harmonic variations of the Earth's magnetic field generated by the solar quiet current vortex in the ionosphere. Interpolation of the MT or magnetic transfer functions is necessary, because the magnetic transfer functions are computed at the frequencies of the Sq harmonics (e.g., 24 h, 12 h, etc), whereas MT transfer functions must be computed at frequencies in the continuum between the Sq *spectral lines* to avoid contamination from the non-uniform Sq source field. Because the Sq source field tends to be highly polarised (see, for example, Figure 10.2), generally only one polarisation of one component of $\underline{C}$ can be determined. Therefore, independent determinations of the static shift for the two polarisations of the MT impedance tensor cannot be obtained. However, if the impedance phases converge at long periods, as is often the case, a 1-D static shift correction relying for example on C_{yx} can be justified.

High-pressure, laboratory measurements on mantle mineral assemblages that are believed to occur at the mid-mantle transition zones (e.g., 410 km) suggest that phase transitions should lead to a sharp decrease in resistivity at these depths (e.g., Xu *et al.*, 1998). Where sufficiently long-period ($\sim 10^5$ s) MT data are available, the hypothesised decrease in resistivity at 410 km can be used during modelling to calculate static shift factors that are in good agreement with those obtained using the equivalence relation given in Equation (5.40) (Simpson, 2002b). For example, if the MT data penetrate deeply enough to indicate a highly conductive layer in the mantle at a depth of 1000 km, then we should be suspicious, and consider the likelihood that a static shift correction is necessary to shift this layer up towards the transition zones. Alternatively, we could perform an inversion with a layer fixed at 410 km and compute the static shift factor that is required directly.

When we choose appropriate methods for correcting static shift, we have to consider the target depth that we are interested in modelling. In complex 3-D environments, near-surface correction techniques may be inadequate if we're interested in long-period data and structure at depth. On the other hand, short-period apparent resistivities may be biased if method (iii) is applied in an environment where data are distorted by deep-seated heterogeneities. Returning to our random-mid-crust model (Figure 5.16), an inductive effect is generated in the period range 20–5000 s, as indicated by impedance phase splitting (Figure 5.21), whereas at longer periods (>5000 s) the impedance phases come together, but the apparent resistivities remain split, and parallel to each other, indicating that

Figure 5.21 (a) Apparent resistivities, and (b) impedance phases synthesised at a site located over the centre of the 'random-mid-crust' model shown in Figure 5.16. Whereas the long-period (>5000 s) impedance phases for the *xy*- and *yx*-polarisations converge, the long-period apparent resistivities are offset parallel to each other owing to galvanic distortion arising from the mid-crustal layer. On the other hand, the impedance phase splitting in the ~20–5000 s period range is indicative of multi-dimensional induction arising from the mid-crustal layer.

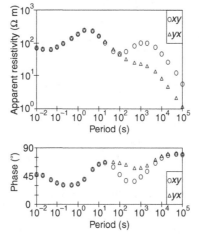

the heterogeneous mid-crustal layer generates galvanic distortion. If we are only interested in determining the depth at which the mid-crustal layer occurs, then the long-period galvanic distortion need not concern us, but if our intended target were to lie deeper than the mid-crustal layer, then the static shift owing to this layer would have to be taken into account (e.g., by placing greater emphasis on modelling the impedance phases). It is good practice to experiment with more than one technique of static shift correction, and to compare how models derived from data subjected to different techniques of static shift correction differ.

5.9 Current channelling and the concept of magnetic distortion

It has been noted in the foregoing sections that the concept of decomposing the impedance tensor into two matrices describing induction and galvanic distortion of electric fields is not a panacea. If we had a facility to model a sufficiently large region of the Earth in a model composed of sufficiently small-scale conductivity heterogeneities, then decomposition might not be necessary. However, a model containing both sufficiently large- and small-scale conductivity structures would demand very high computational power. Furthermore, information about the small-scale conductivity heterogeneities responsible for distortion is generally not available or, in other words (as discussed in Section 5.3), the decomposition model is under-parameterised.

Current channelling (concentration of induced currents in highly conductive elongated structures) also causes distortion of geomagnetic variational fields, which can be treated either as a decomposition or a 3-D modelling problem. Jones (1983) reviews early work on current channelling and points out that channelling appears to occur if models that are too small to include the region where the channelled currents are induced are considered. This is similar to the argument relating to distortion of the impedance tensor – that the modelling approach requires a single model that incorporates a sufficiently large region of the Earth on the one hand, and sufficiently small-scale conductivity anomalies on the other. Two solutions for treating distortion of geomagnetic variational fields have been proposed:

(i) The calculation of a 'current channelling number', N_{chan}, or a dimensionless ratio of the current in the elongated structure to the total available current (Edwards and Nabighian, 1981; Jones, 1983). Such a ratio should depend on the conductivity contrast, $\sigma_{\text{chan}}/\sigma_{\text{host}}$, and the ratio of the channel volume, V_{chan}, to the volume, p^3_{host}, of the inductive space, e.g.,

$$N_{\text{chan}} = \frac{\sigma_{\text{chan}}}{\sigma_{\text{host}}} \frac{V_{\text{chan}}}{p^3_{\text{host}}}. \tag{5.41}$$

Equation (5.41) provides only a qualitative answer as to whether current channelling does or does not occur. A more quantitative answer should also incorporate the ratio of the distance between the measurement site and the channel divided by the penetration depth, as suggested by the Biot–Savart Law. Current channelling numbers indicate that the effect of magnetic distortion created by a single channel decreases with increasing period (and penetration depth).

(ii) Measured data used to calculate the *tipper* matrix (Section 2.8) can be decomposed into two matrices describing induction and galvanic distortion of magnetic fields (Zhang *et al.*, 1993; Ritter and Banks, 1998). The anomalous magnetic field generated by the channelling is in phase with the regional electric field. Therefore the definition of the tipper matrix (Equation (2.49)) is modified to:

$$(T_x, T_y) = [(T_x^{\,0}, T_y^{\,0}) + (Q_{zx}, Q_{zy})Z](I + Q_{\text{h}}Z^0), \tag{5.42}$$

where (T_x, T_y) and (T_x^0, T_y^0) are the total and regional *tipper vectors*, respectively. (Q_{zx}, Q_{zy}) is the response of a distorting body acting on the regional horizontal electric field to produce the current channelling effect on the vertical magnetic field (Zhang *et al.*, 1993). The term $Q_{\text{h}}Z^0$ describes the distortion of the horizontal magnetic field created by the current channelling. The appearance of the regional impedance tensor in Equation (5.42) allowed Ritter and Banks (1998) to use tipper data from the Iapetus array of magnetovariational data to extract a regional strike of the conductivity structure without using MT techniques.

Chave and Smith (1994) extend the decomposition of the magnetotelluric impedance tensor (Equation (5.22)) by considering the distortion of the horizontal magnetic field created by channelling of both electric and magnetic fields.

Chapter 6
Numerical forward modelling

Suppose that after applying some of the tests suggested in the last chapter, we have computed 'undistorted' transfer functions and we infer that for a part of the target area (or for a limited period range) a 2-D interpretation of the conductivity distribution may be adequate, but that treating the target area in its entirety requires a 3-D interpretation. This is exactly the information that is required now, in order to decide which numerical modelling scheme should be applied for the purpose of generating synthetic transfer functions that can be compared with the measured ones. Most modelling schemes divide a 1-, 2- or 3-D world into blocks or cells and allow the user to choose the conductivity within each individual cell. Maxwell's equations are then solved within each cell, while certain boundary and continuity conditions are observed. Here we restrict ourselves to practical hints for those users who intend to apply existing modelling schemes to their data. There is an excellent book on the subject (Weaver, 1994) for those who wish to write their own modelling code. Some words of caution: choosing the conductivity inside some cells is somewhat arbitrary, because a change does not affect the calculated transfer functions significantly. Modelling experts explore the relationship between model changes and result changes with 'sensitivities'. Obviously, more than one realisation of the model has to be computed before a small misfit (a normalised difference between field transfer functions and the modelled ones) is found. Users who don't like the trial-and-error game described here might proceed to the inversion, but see our warnings in Chapter 7!

6.1 Why forward modelling, and which scheme?

Forward modelling of MT data involves simulation of the electromagnetic induction process with a computer program. For example, if the model parameters of a 1-D Earth – the layer thicknesses and conductivities – are known, then with Equation (2.34) and the iterative use of Equation (2.33), the *Schmucker–Weidelt transfer function* at the surface can be computed. Some readers might think that this is not really what they are aiming for. After all, they *have* the data at the surface of the Earth – for example, the magneto-telluric impedances from the processing of field data – and it is the model parameters that they seek to obtain. In other words, the desired result is the inverse of the simulated induction process. However, before you move on to the next chapter, 'Inversion of MT data' consider the following reasons for doing forward modelling.

1 For those who are planning a field campaign, or designing an instrument, it may be instructive to synthesise data using a simple model, or a group of simple models (see, for example, Section 3.1.1). Such an exercise can provide answers to questions of the type: 'What periods are required to resolve a conductivity anomaly in the so-and-so depth range?'; 'If sites are installed at a given distance from a vertical boundary (e.g., a land – ocean transition) will the MT responses be perturbed, and in which period range?'. Depending on the answers to these questions, it may prove necessary to modify the design of the field campaign, or of the instrument. Forward modelling is a cheap and relatively simple exercise compared to the input (human and fiscal) involved in re-doing a field campaign, or re-designing an instrument.

2 The benefits of forward modelling also extend to those who already have a dataset from which they wish to generate a model. There may be no inverse scheme available that is adaptable to their site layout, or to 3-D effects in their data, and even in straightforward cases, where an inverse scheme is available, forward modelling can be used to investigate whether features included in a model obtained by inversion are constrained (or even required) by the input data. For example, by applying a forward-modelling scheme on models with and without particular conductivity structures, the difference between response functions for two models can be assessed, and compared with the errors in the field data. If the responses for both models lie within the *confidence intervals* of the data, then the conductivity structure in question cannot be said to be required by the data. In Section 7.3 ('Artefacts of inversion'), we give a more specific example of the benefit of combining forward modelling and inversion.

When choosing a forward-modelling algorithm, the most important question is: 'Do we want to model a 1-D, a 2-D or a 3-D Earth?' – geophysics jargon for 'Do we want to simulate the induction process in a layered Earth, a 2-D Earth in which the conductivity varies only in the vertical and in one horizontal direction, or a 3-D Earth?' The real Earth is 3-D, but 3-D codes require appreciable

computational power, long computational times and, more import-
antly, using them can be time-consuming for the user. If the modelling is related to field data, the tests suggested in the last chapter can
be applied to find out whether the regional conductivity distribution
is 1-D (Equation (5.21)) or 2-D (Equation (5.32)), to some extent.
Often, the answer will depend on the period range, because the
dimensionality measures are period dependent. In the past, 2-D
forward modelling and, more often, 2-D inversion has been used
to explain field data from a 3-D field environment for at least one of
the following reasons.

1 No 3-D code was available that could adequately handle the theoretical
  requirements (e.g., the gridding rules outlined in Section 6.2) and the size
  of the target area.
2 The dimensionality measures were not available or the tests have not been
  performed.
3 The magnetotelluric method was either used in conjunction with seismic
  reflection, or there was an intention to combine the two methods.
  Seismic reflection is performed along profiles and the results are presented in vertical cuts through the Earth, just as when displaying a 2-D
  model.
4 The magnetotelluric measurements themselves were conducted along a
  profile, and even though dimensionality tests might indicate a 3-D Earth,
  a 3-D model cannot be adequately constrained, because there are no data
  from off the profile.

We believe that (1) will occur less often, (2) does not need to occur,
(3) will probably occur less often, either because of the use of 3-D
seismics, or because joint interpretations of MT and seismic data are
performed less routinely (see Chapter 9.1), and (4) will, hopefully,
occur less often. There may still be field situations where we have no
other choice than to install sites along a single profile (e.g., if there is
only one road through a mountainous region). However, it has been
demonstrated (Simpson, 2000) that, even when logistical constraints
dictate non-ideal site distributions, a 3-D model can provide some
insight into the geometry of conductors.

6.2 Practical numerical modelling: gridding rules, boundary conditions, and misfits revisited

6.2.1 Gridding rules and boundary conditions

We refer the reader to Weaver (1994) for a summary of work done on
2-D modelling. With respect to 3-D modelling, differential equation

methods and integral equation methods have been proposed. Whereas in differential equation methods both the conductivity anomaly and the surrounding 'normal world' is covered with a network of cells, in the integral equation technique only the anomalous conductivity structure is 'gridded'. The latter requires less computational power and it is probably for this reason that most of the early contributions were devoted to the integral equation method (e.g., Raiche, 1974; Hohmann, 1975; Weidelt, 1975; Wannamaker et al., 1984b). Considering only the region with anomalous conductivity is made possible by a volume integration of Maxwell's equations in conjunction with Green's theorems (Zhdanov et al., 1997). The electromagnetic field in the model is described by the Fredholm integral equation:

$$\mathbf{U}(r) = \mathbf{U}^n(r) + \int_{V^a} \underline{G}(r/r') \, \mathbf{j}(r') \, dV' \tag{6.1}$$

for the vector field $\mathbf{U}$ (either the electric or the magnetic field). Here, $\underline{G}$ is Green's tensor, $\mathbf{j} = \sigma^a \mathbf{E}$ is the current density owing to the anomalous conductivity, σ^a, and V^a is the volume of the region with anomalous conductivity. For the numerical solution, V^a is subdivided into cells in which the functions $\mathbf{U}(r)$ and $\sigma^a(r)$ are assumed to be constant. Hence, the problem is reduced to the solution of a linear system of equations, in which the integral in Equation (6.1) is replaced by a sum.

Integral equation methods have rarely been used for interpreting measured electromagnetic data, probably because the model of an isolated 'anomaly' surrounded by a layered *half-space* generally does not describe realistic tectonic situations. In contrast, differential equation methods allow for the computation of electric and magnetic fields in arbitrarily complex conductivity structures, but require more computational power (both storage space and computing time). Using these methods, a boundary value problem is solved either with finite differences (FD) or with finite elements (FE). With few exceptions (e.g., the FE code by Reddy et al., 1977) most applications of differential equation methods to EM rely on FD techniques. Here, the linear relations between the fields at the nodes of a 3-D grid are of interest. In the early FD techniques, differential equations – either the Maxwell equations (Equation (2.6)) or the induction equations (Equations (2.14) and (2.15)) – were replaced by finite differences. For example,

$$\frac{dU}{dx} = \frac{U(x_2, y, z) - U(x_1, y, z)}{\Delta x}, \tag{6.2}$$

where U is any field component and Δx is the distance between neighbouring nodes (x_1, y, z) and (x_2, y, z). For example, the y-component of Equation (2.6b):

$$\frac{\mathrm{d}B_x}{\mathrm{d}z} - \frac{\mathrm{d}B_z}{\mathrm{d}x} = \mu_0 \sigma E_y$$

is replaced by

$$\frac{B_x(x,y,z_2) - B_x(x,y,z_1)}{\Delta z} - \frac{B_z(x_2,y,z) - B_z(x_1,y,z)}{\Delta x} = \mu_0 \sigma E_y(x,y,z).$$

(6.3)

Modern FD applications (Mackie *et al.*, 1993; Mackie and Madden, 1993) use the integral form of Maxwell's equations. For example, the integral form of Ampère's Law:

$$\oint \mathbf{B}\mathrm{d}l = \int\int \sigma \mathbf{E}\mathrm{d}s$$

(6.4)

is translated into a difference equation (instead of Equation (2.6b)). Similarly, the integral form of Faraday's Law:

$$\oint \mathbf{E}\mathrm{d}l = \int\int i\omega \mathbf{B}\mathrm{d}s$$

(6.5)

is the integral formula equivalent to Equation (2.6a). The numerical realisation of discrete spatial differences (of which Equation (6.2) is the simplest example) is replaced by a numerical averaging that simulates the integrals in Equations (6.4) and (6.5).

Thin-sheet modelling takes advantage of the inherent ambiguity involved in distinguishing between the conductivity and thickness of conductivity anomalies by squashing anomalous conductances into an infinitesimally thin layer (Figure 6.1). This approach reduces the computational problem from one of calculating perturbations in two parameters – conductivity and thickness – to one of calculating perturbations in one parameter – conductance. For example, the thin-sheet modelling algorithm of Vasseur and Weidelt (1977) allows for a 3-D structure within one layer of an otherwise layered Earth. Thin-sheet modelling algorithms have been successfully applied in studies involving land–ocean interfaces and seawater bathymetry when the targeted *electromagnetic skin depth* is greater than the seawater depth (see Section 9.4). Wang and Lilley (1999) developed an iterative approach to solving the 3-D thin-sheet inverse problem, and applied their scheme to a large-scale array of geomagnetic induction data from the Australian continent.

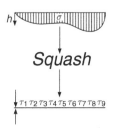

Figure 6.1 Illustration of the concept of thin-sheet modelling: a conductivity anomaly with conductivity, σ, and variable thickness is represented within an infinitesimally thin sheet composed of discretised conductances (τ_1, τ_2, τ_3, etc).

An early application of 3-D FD modelling (Jones and Pascoe, 1972) by Lines and Jones (1973) illustrates how sign changes of components of the electromagnetic fields with periods of order 30 minutes might arise from island effects. More general applications of 3-D FD modelling techniques to tectonic and geodynamic problems (e.g., Masero *et al.*, 1997; Simpson and Warner, 1998; Simpson, 2000, Leibecker *et al.*, 2002) came later than thin-sheet model studies, one reason for this tardiness being the computational power required if all rules for the design of a 3-D grid are followed properly.

Weaver (1994) provides some practical rules for the design of a grid for 2-D FD forward modelling, which can, with some modifications, also be applied to 3-D FD modelling.

'The following guidelines . . . have been found helpful in designing a satisfactory grid

1 Up to at least two electromagnetic skin depths on either side of a vertical or horizontal boundary separating two regions of different conductivity, the node separation should be no more (and very close to the boundary preferably less) than one-quarter of an electromagnetic skin depth. At greater distances where the field gradient will be smaller, this condition can be relaxed.

2 The first and last (*M*th) vertical grid lines should be placed at least three electromagnetic skin depths beyond the nearest vertical conductivity boundary. Here 'electromagnetic skin depth' refers to the electromagnetic skin depth in the most resistive layer of the relevant 1-D structure.

3 The separation of adjacent nodes should be kept as nearly equal as possible near conductivity boundaries, and preferably exactly equal across the boundaries themselves. When a rapid change in the node separation from one part of the model to another is required, the transition should be made as smoothly as possible, preferably in a region where field gradients are expected to be small and with no more than a doubling (or halving) of adjacent separations.'

Computing model data for different periods will result in different electromagnetic skin depths being realised. ('Two electromagnetic skin depths' is a larger distance at long periods than at short periods). Weaver's guideline 1 implies that the node separation can be increased with increasing distance from a boundary. However, when we increase node separations, we must take care not to violate guideline 3. In order to follow guidelines 1 and 3, we might be forced to choose small (1/4 of the smallest electromagnetic skin depth) node separations over a very wide area. Weaver's guideline 2 is a practical consequence of the 'adjustment length' concept (Section 2.7): a vertical conductivity boundary can influence the electromagnetic field as far away as two to three electromagnetic skin depths. This guideline has its equivalent for 3-D FD modelling. Horizontal and vertical cross-sections through a grid employed for 3-D FD modelling (Mackie *et al.*, 1993) of an array of 64 MT sites stationed over a 300×400 km^2 area in Germany are shown in Figure 6.2. In contrast, in integral equation methods, the grid only covers the anomalous domain (e.g., Raiche, 1974).

Designing a grid involves a trade-off between computational economy and precision. A practical way to explore the effect of node separation and the model size on the modelled data could be the following: for the same model, we start by designing a grid where the above guidelines are observed to some extent; we subsequently decrease the node separation and increase the number of nodes. If the model data do not change, then our starting grid was already fine enough. On the other hand, a perturbation of the model data indicates that a finer grid is required, and the sequence should be repeated.

6.2.2 Misfits revisited

In Section 5.3, we introduced misfit measures describing the applicability of general models (e.g., the layered-Earth model or the 2-D/3-D *superimposition model*) to a particular dataset. More generally, a misfit is a difference between measured and modelled data, e.g.,:

$$\varepsilon^2 = \sum_{j=1}^{jsites} \sum_{i=1}^{ifreq} \frac{\left| C_{ij} - C_{ij,\, \text{mod}} \right|^2}{\left| \Delta C_{ij} \right|^2}, \tag{6.6}$$

where C_{ij}, $C_{ij,\text{mod}}$ and ΔC_{ij} are the measured and modelled *transfer functions* at site j and evaluation frequency i and the *confidence interval* of the measured data, respectively. Of course, for the case of 1-D models, only one site is evaluated.

Is it desirable to have a very small misfit? Too small a misfit can mean that the model resolves data noise: suppose we have 10 sites and 20 discretised frequencies, then the misfit given in Equation (6.6) should not be smaller than 200, because for a misfit smaller than 200 the average distance between measured and modelled data will be smaller then the confidence interval. However, practical experience tells us that there is rarely a model with such a small

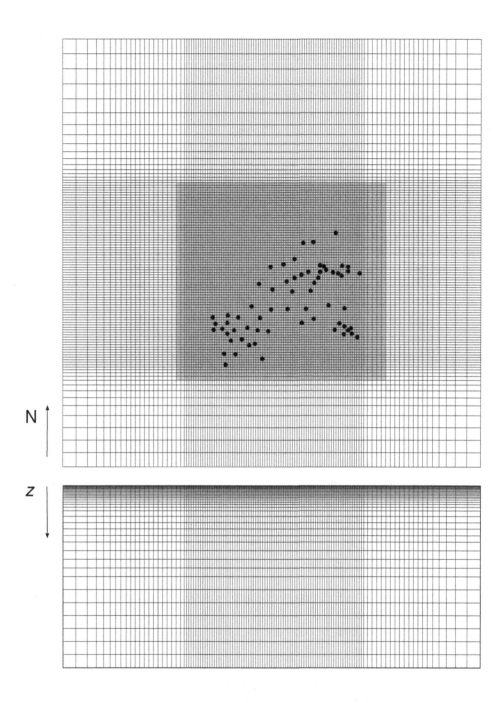

Figure 6.2 Example of a grid used in 3-D FD modelling. The horizontal cross-section (*top*) consists of 103×119 cells and represents a geographical area of 858×923 km² in Germany. The black dots indicate the locations of MT sites. These lie within the core of the grid, where cell dimensions are 5×5 km². The vertical cross-section (*bottom*) contains 34 layers that extend to a depth of 420 km. Boundary conditions are imposed by a 1-D background model. Overall, the grid should be envisaged as a cube with 103×119×34 cells. (Courtesy of A. Gatzemeier.)

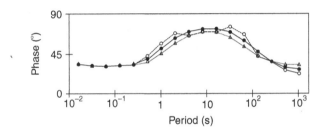

Figure 6.3 The two impedance phase curves plotted as open symbols and dashed lines have the same ε^2 misfits with respect to the impedance phase curve represented by the closed circles and solid line, but the forms of these curves are different. Which of these data fits is preferable?

misfit. For 1-D and 2-D modelling, this is so because the modelled data are rarely truly 1-D or 2-D: even if the dimensionality criteria discussed in Section 5.3 suggest a 3-D model, MT practitioners often use 2-D modelling.

Even if 3-D modelling is performed, the data will often not support a quest for the best-fit 3-D model, because the area represented by the 3-D model is not covered densely enough with MT sites. Is 3-D modelling useful at all in this situation? Yes, if we use forward modelling for *hypothesis testing*: by running different models, we can assess whether a particular feature in a model is really **required** by a particular dataset – if so, the misfit will be lowered by inclusion of the feature in question. The trade-off between complicated and rough models, which generate a small misfit, and 'smooth' models, which generate larger misfits, will be discussed in Section 7.2.

The misfit parameter expressed in Equation (6.6) is an arbitrary mathematical quantity, and because it is possible to derive models that have the same misfits, but which contain more or less structure than is present in the data, it is worthwhile to compare the shapes of the modelled and measured transfer functions (see, for example, Figure 6.3).

6.3 From computed electric and magnetic fields to transfer functions

Some forward-modelling algorithms output electric and magnetic field responses at the surface of the model, rather than transfer functions. The responses are computed for two different (generally orthogonal) polarisations of the regional field, so that a set of ten model fields is required at each period and computation site in order to represent the five electromagnetic field components – E_x, E_y, H_x, H_y, H_z. If you understand why it is necessary to compute the responses for two different polarisations of the regional field, then you will understand how to compute transfer functions from the surface electric and magnetic fields in the output model files. In Chapter 4, the ideas of auto- and cross-spectral correlation were introduced as a means of extracting the *impedance tensor* elements from noisy data. For synthetic data, we simply need to compute the response for two different orientations of the regional field (or equivalently two different orientations of the model) in order to be able to solve a system of simultaneous equations. For example, with subscripts 1 and 2 denoting the two orientations of the regional fields we obtain, from Equation (4.14a),:

$$E_{x_1}(\omega) = Z_{xx}(\omega)H_{x_1}(\omega) + Z_{xy}(\omega)H_{y_1}(\omega) \tag{6.7a}$$

$$E_{x_2}(\omega) = Z_{xx}(\omega)H_{x_2}(\omega) + Z_{xy}(\omega)H_{y_2}(\omega). \tag{6.7b}$$

Dividing Equation (6.7a) by $H_{x_1}(\omega)$ and Equation (6.7b) by $H_{x_2}(\omega)$, and subtracting Equation (6.7b) from Equation (6.7a) gives:

$$\frac{E_{x_1}(\omega)}{H_{x_1}(\omega)} - \frac{E_{x_2}(\omega)}{H_{x_2}(\omega)} = Z_{xy}(\omega)\left(\frac{H_{y_1}(\omega)}{H_{x_1}(\omega)} - \frac{H_{y_2}(\omega)}{H_{x_2}(\omega)}\right), \tag{6.8}$$

so

$$Z_{xy}(\omega) = \frac{E_{x_1}(\omega)H_{x_2}(\omega) - E_{x_2}(\omega)H_{x_1}(\omega)}{H_{y_1}(\omega)H_{x_2}(\omega) - H_{y_2}(\omega)H_{x_1}(\omega)}. \tag{6.9}$$

More generally, for a bivariate linear system with input processes X and Y, output process Z and transfer functions A, B:

$$\left. \begin{aligned} Z &= AX + BY \\ A &= \frac{Z_1 Y_2 - Z_2 Y_1}{X_1 Y_2 - X_2 Y_1} \\ B &= \frac{Z_2 X_1 - Z_1 X_2}{X_1 Y_2 - X_2 Y_1} \end{aligned} \right\}. \tag{6.10}$$

By making appropriate substitutions in Equation (6.10), all components of the impedance tensor and *tipper vector* can be computed.

6.4 Avoiding common mistakes

Too complicated a start model.
Suppose the dimensionality analysis suggests a 3-D situation, and a 3-D code is selected. Even if the data indicate the existence of many different conductivity structures, one should not start with a complicated model. A better approach is to start with a simple model, and observe how the modelled transfer functions develop during the process of complicating the model.

Over-interpretation of transfer function amplitudes.
Because of the 'distortion' described in Chapter 5, the amplitudes of magnetotelluric transfer functions from field data can be poorly resolved. We therefore suggest to reproduce the impedance phase data with forward-modelling schemes. Geomagnetic transfer functions (*induction vectors*) can also be distorted by *galvanic effects* (Section 5.9), and this possibility should be considered when modelling these transfer functions with a code that simulates the induction process. Theoretically, if large- and small-scale structures are included in the same model and if the model is large enough, the galvanic effects in the modelled transfer functions can be properly described.

MT data constrain conductances rather than resistivities and thicknesses of conductors.
Discriminating between a thick, moderately conductive anomaly and a thin, highly conductive anomaly is possible only with high-quality MT data acquired in regions of low geological noise. Therefore, the thickness of a conductor is generally poorly constrained. The constraints (or lack thereof) on the thickness of a modelled conductivity anomaly can easily be investigated using forward modelling, by redistributing the conductance of a modelled anomaly into a thicker or thinner zone, and comparing the misfits that result from the different models. The proximity of apparent resistivity and impedance phase curves for different distributions (variable resistivities and thicknesses) of a 1000 S mid-crustal conductor is illustrated for a 1-D model in Figure 6.4 and for a 2-D

Figure 6.4 Comparison of apparent resistivity and impedance phase curves for three 1-D models in which a 1000 S mid-crustal conductance is distributed in layers of 10 km × 10 Ω m, 5 km × 5 Ω m and 1 km × 1 Ω m, respectively. The largest impedance phase difference of (∼6°) for a 1-km-thick, 1 Ω m conductor versus a 10-km-thick, 10 Ω m conductor occurs at 4.1 s.

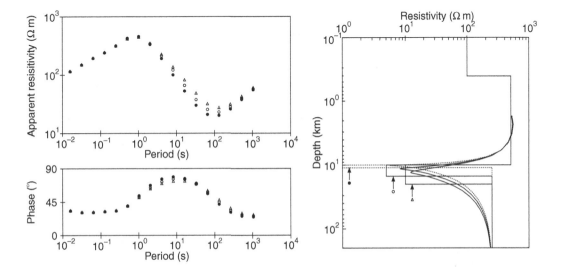

model in Figure 6.5. The problem of distinguishing between thick and thin conductors is exacerbated if the overburden conductance is increased or the conductance of the anomaly is decreased.

The generalised model space might be too small.

This can lead to problems when attempting to model impedance phases (and apparent resistivities) and *induction arrows* jointly. Induction arrows are often influenced significantly by regional conductivity anomalies that lie outside the MT model space. These different data sensitivities can be manifest as a lack of orthogonality between *phase-sensitive strike* and induction arrow orientation. The discrepancy can arise owing to current channelling, and can be modelled if the model space is sufficiently large and sufficiently finely gridded.

Being hooked on a band-wagon.

We should be aware of the possibility that we do have prejudices (although being open-minded people, otherwise). Here is an example: we originally assume that conductivity is isotropic and try to interpret a 2-D dataset with a 2-D model. To reproduce the measured impedance phase data from both polarisations, we run the 2-D code twice – once for the **E**-*polarisation* and once for the **B**-*polarisation* (remember from Section 2.6 that the equations for the two polarisations are decoupled and therefore they can be treated independently). The two modes produce models that differ significantly in places, and attempts to model both modes jointly result in models with larger misfits. A possible solution could be to allow for anisotropic conductivity in those parts of the model that differ. The lesson here is that a large misfit is not always a problem, but can be a challenge. If we rise to this challenge, then we may re-think the underlying assumptions in our model, rather than introducing or removing a few conductive blobs here and there. Of course, the opposite effect can also occur. For example, we are so much taken away with the concept of anisotropy that we interpret impedance phase differences between the two polarisations always with anisotropic models. The possibility that other features in the model are responsible for impedance phase differences must not be neglected!

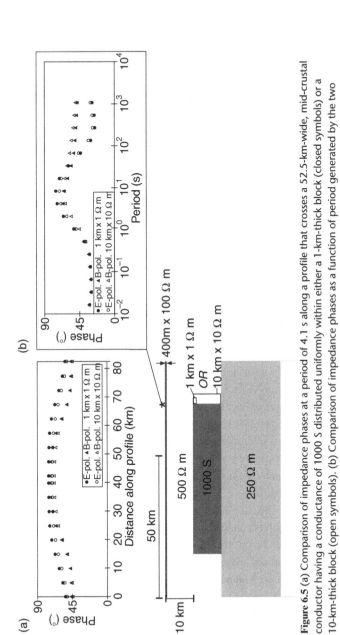

Figure 6.5 (a) Comparison of impedance phases at a period of 4.1 s along a profile that crosses a 52.5-km-wide, mid-crustal conductor having a conductance of 1000 S distributed uniformly within either a 1-km-thick block (closed symbols) or a 10-km-thick block (open symbols). (b) Comparison of impedance phases as a function of period generated by the two conductivity–thickness distributions in (a), for a site (*) located over the mid-crustal block.

Chapter 7
Inversion of MT data

Inversion schemes provide a 'fast-track' means for modelling data without the grind of forward modelling by allowing you to go from the data to the model, rather than the other way round. But beware – there are pitfalls, especially for the inexperienced!

In some inversion schemes the non-linear relationship between the model and the data is forced to be 'quasi-linear' by use of a transformation into a substitute model and data. In others, small model changes are assumed to have a linear relationship to small data changes. Once the required linear system of equations has been derived, the problem is reduced to one of inverting a matrix. The 'Monte-Carlo inversions' form another class of modelling algorithm. From a mathematical point of view, they are not strictly inversions at all, but rather long sequences of forward models in which conductivities are perturbed at random – as in a casino (à la Monte Carlo).

Because electromagnetic energy propagates diffusively, MT sounding resolves conductivity gradients, rather than sharp boundaries or thin layers. We should, therefore, think of MT as producing a blurred rather than a focussed image of the real Earth structure. Furthermore, because MT sounding is a volume sounding, the same features will be sampled by MT data at neighbouring sites when the inductive scale length is of the same order as or greater than the site spacing. Many inversion schemes are, therefore, founded on the philosophy of minimising model complexity, wherein, rather than fitting the experimental data as well as possible (which maximises the roughness of the model), the smoothest model that fits the data to within an accepted tolerance threshold is sought. On the one hand, 'smooth' or 'least-structure' models reduce the temptation to over-interpret data. On the other hand, sharp boundaries do occur quite often in nature, an example being the contact between an ore body and its resistive host rock.

Although a good inversion scheme can provide the model with the smallest misfit, you should have some experience with forward modelling before you apply an inverse scheme: suppose you do a 1-D (layered half-space) inversion and find a conductive layer at some depth, which you excitedly proceed to interpret. A quick check with a 1-D forward scheme might have shown you that this layer is the product of a single data point. In other words, forward modelling is a good tool to explore the constraints provided by your data.

Many practitioners are content to compute one best-fit model for their data. Always be aware of the inherent non-uniqueness associated with modelling, and explore more than one model. Forward modelling is a good way of exploring allowable perturbations to a model output from an inversion. What happens, for example, if your target area includes an electrically anisotropic domain, but your inverse scheme doesn't allow for that possibility? Ironically, a lot of support for the anisotropy concept came from applications of 2-D inversions: applying the inversion to the transfer functions derived from electric fields in two perpendicular directions often produces two contradictory models, both of which have small misfit measures, indicating very different conductivities in a particular depth range. Again beware! Anisotropy is certainly the easiest way to fit MT data for which the apparent resistivities and phases of the two principal polarisations differ, but there must be a full justification for preferring anisotropy to lateral effects.

Unfortunately, many users apply inverse schemes only because of their availability, in which cases even if the tests mentioned in Chapter 5 suggest a 3-D situation, a 2-D inversion is used because a 3-D inversion scheme is not available. On the other hand, 3-D modelling is computationally viable using forward-modelling algorithms.

7.1 Forward, Monte Carlo and inverse modelling schemes

Forward modelling is an iterative procedure involving progressive trial- and-error fitting of data by (i) computing the responses of an input model, (ii) comparing these with measured data, (iii) modifying the model where the data are poorly fitted and then (iv) re-computing the responses, until a satisfactory fit to the measured data is achieved. This can be a tedious process.

Although sometimes (incorrectly) referred to as an inverse scheme, the *Monte-Carlo technique* is an automated form of forward

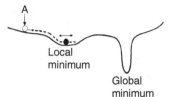

Figure 7.1 Simple analogy of a ball-bearing released at position A falling into a local minimum, where it becomes trapped before it can reach the global minimum of the surface along which it rolls.

modelling in which misfits between model data and measured data are calculated for a sequence of models, to which random changes are made, until the misfit falls below a specified threshold, or further perturbations do not produce a reduction in the misfit, suggesting that a minimum has been reached. A common misfit measure used in modelling is the root-mean-square (rms) relative error of fit, ε, as defined by Equation (6.6). Because of the randomness of the technique, different initial models can lead to different final models that fit measured data equally well. This allows the range of models that can fit a measured dataset to be explored relatively quickly. However, sometimes the minimum misfit attained between modelled and measured data using the Monte-Carlo technique may be a local minimum (Figure 7.1), making further investigation with forward modelling advisable. Models derived using the Monte-Carlo technique can be used as initial inputs in forward-modelling or inversion schemes.

As the name suggests, inversion allows the user to progress directly from data to a model, rather than the other way round. As with forward modelling, inversion is an iterative process, the aim of which is to progressively reduce the misfit between measured data and data synthesised from a model. All inversion problems require the ability to solve the forward problem as a pre-requisite. The inverse problem that links a set of MT responses to a conductivity model is non-linear. However, the non-linear relationship between the model and the data can often be adequately approximated as quasi-linear by expanding the MT response, R, for a conductivity model, $\sigma(z)$, at period, T, about an arbitrary starting model, $\sigma_0(z)$, according to:

$$R(\sigma, T) = R(\sigma_0 + \delta\sigma, T) = R(\sigma_0, T) + \int\limits_0^\infty G(\sigma_0, T, z)\delta\sigma(z)\mathrm{d}z + R_\sigma, \quad (7.1)$$

where R_σ represents the remainder term for the first-order expansion in conductivity and G is the Fréchet kernel (assuming that R is Fréchet differentiable). Assuming that the remainder term is second order in $\delta\sigma$, and can therefore be neglected, Equation (7.1) can be expanded as an approximate linear equation that relates changes in the modelled responses to a linear functional of the conductivity model (e.g., Dosso and Oldenburg, 1991):

$$\delta R(\sigma, \sigma_0, T) + \int_0^\infty G(\sigma_0, T, z)\sigma_0(z)\mathrm{d}z = \int_0^\infty G(\sigma_0, T, z)\sigma(z)\mathrm{d}z, \qquad (7.2)$$

where

$$\delta R(\sigma, \sigma_0, T) = R(\sigma, T) - R(\sigma_0, T)$$

and

$$\delta\sigma(z) = \sigma(z) - \sigma_0(z).$$

The problem of finding a model that minimises the misfit between measured and modelled data is now reduced to one of inverting a matrix, for which a number of standard methods exist. Because higher-order terms in the expansion are neglected, the procedure must be applied iteratively until an acceptable model is realised.

Of the plethora of 1-D inverse schemes proposed in the geophysical literature, conductance delta functions (Parker, 1980; Parker and Whaler, 1981) represent one extreme, and *least-structure* inverse schemes (e.g., Constable *et al.*, 1987) the other, with layered models (e.g., Fischer and LeQuang, 1982) occupying a niche in between. Most 2-D inversion schemes incorporate the least-structure philosophy, which we shall explore in more detail in the next section. Some 2-D inversion schemes allow intrinsic anisotropy to be incorporated into the model (e.g., Pek and Verner, 1997). 3-D inversion schemes are still uncommon. Iterative solutions to the 3-D inverse problem have been attempted by Wang and Lilley (1999) and Hautot *et al.* (2000). However, input data from array studies is needed if 3-D inversion is to provide meaningful models.

7.2 D^+ optimisation modelling versus least-structure philosophy, or a treatise on the difference between the mathematical and the physical viewpoint

Many inversion schemes are founded on linearisation about an arbitrary starting model of an inherently non-linear system of equations, as described in the previous section. However, Parker (1980) presented a non-linear formulation whereby the optimal 1-D solution is described by a stack of delta functions (infinitesimally thin sheets of finite conductance) embedded in a perfectly insulating *half-space*. This is known as the D^+ *model*. The fit of the modelled D^+ data to the measured data can be expressed in terms of the weighted least-square statistic given in Equation (6.6), with *jsites*=1. It should be noted that this ε^2 misfit statistic depends, in part, on the distribution of errors in the measured data. Because of measurement errors and inevitable departures from the theoretically idealised 1-D dimensional sounding environment, ε^2 is not expected to vanish to zero. The expected value of ε^2 for data with independent Gaussian errors is *ifreq*. If models whose misfits are less than two standard deviations of ε^2 above its expected value are deemed acceptable, then this defines a misfit inequality (cf. Parker and Whaler, 1981):

$$\varepsilon^2 < ifreq + 2\sqrt{2(ifreq)}. \tag{7.3}$$

Failure to satisfy this inequality implies that data cannot be satisfactorily modelled following the assumption of a 1-D Earth, and that multi-dimensional conductivity structures must be considered.

In limited situations, the calculation of D^+ solutions may represent a useful initial stage to modelling, offering confirmation, or otherwise, as to the existence of 1-D solutions. However, such sharp interfaces as delta functions are physically unrealistic. Parker and Whaler (1981), therefore, developed the H^+ *model*, in which the best-fit to a model composed of a stack of homogeneous layers overlying a perfect conductor placed at the maximum *penetration depth* is sought. However, depending on the error structure of measured data, approaches to modelling aimed solely at minimising the misfit between modelled and measured responses are not necessarily justified. Fitting electromagnetic data as closely as possible maximises the *roughness* of a model, and it is generally desirable to sacrifice a degree of fit to a dataset in favour of introducing gradational conductivity contrasts (Constable *et al.*, 1987).

The conductivity structure apparent in models derived from layered 1-D inversion schemes can depend quite critically on an often arbitrarily pre-determined number of layers that should be present in the final model. Under-parameterisation of the inverse problem tends to suppress structure that could be significant, whilst over-parameterisation introduces structure that is redundant, and not truly resolvable by the available data (Constable *et al.*, 1987). Also, models with too many layers may develop oscillations as the modelled structure tends to the delta function (D^+) solution that optimises the misfit. One solution to this problem is the minimum layer inversion (Fischer *et al.* 1981), in which additional layers are introduced into a model at progressively greater depths only when successively longer-period data demands their presence. On the

other hand, Constable *et al.* (1987), drawing on the philosophy
embodied in the tenet known as Occam's razor[10], propose finding
the smoothest possible model (known as the *least-structure model*)
consistent with an acceptable (user-definable) fit to the data. Given
the diffusive nature of electromagnetic fields, passive EM techniques resolve conductivity gradients rather than sharp boundaries at depth, giving least-structure models a certain appeal. By
explicitly minimising structural discontinuities, unjustifiably complex interpretations of data are avoided, and any structure present
within the model should lie within the resolving power of the data.
Therefore, the temptation of over-interpreting data is reduced.
However, as we shall see in Chapters 8 and 9, in certain depth ranges
such as at the lithosphere–asthenosphere boundary and at the
410 km *transition zone*, the existence of discontinuities is widely
expected. Therefore, layered models and least-structure (smooth)
models represent two extremes between which the resistivity–depth
distribution of the real Earth is likely to lie.

2-D models are constructed on a rectangular grid of rows and
columns intersecting at nodes in the y–z co-ordinate plane (as, for
example, delineated in Figure 2.5) to form blocks or cells, each of
which is attributed a uniform conductivity. Rules of thumb for
constructing such a mesh are dealt with in Section 6.2. Once defined,
the mesh remains fixed from one iteration to the next. Most
2-D inversion algorithms (e.g., *Occam Inversion* (de Groot-Hedlin
and Constable, 1990); *Rapid Relaxation Inversion (RRI)* (Smith and
Booker, 1991)) are founded on the least-structure philosophy,
and involve joint minimisation of data misfit (e.g., expressed as an
rms statistic as given in Equation (6.6)) and model roughness. For
1-D models, roughness can be defined as

$$M_{R1} = \int \left(\frac{dm(z)}{dz} \right)^2 dz \tag{7.4}$$

or

$$M_{R2} = \int \left(\frac{d^2 m(z)}{dz^2} \right)^2 dz \tag{7.5}$$

where $m(z)$ is either conductivity, σ, or $\ln(\sigma)$ (Constable *et al.*, 1987).
In 2-D models, both horizontal and vertical conductivity gradients
can be minimised (e.g., by minimising conductivity differences

---

[10] Named after William of Ockham, a fourteenth-century Franciscan monk who
wrote 'Plutias non est ponenda sine necessitate'. (Entities should not be
multiplied unnecessarily.)

between laterally and vertically adjacent cells in the model). This process is sometimes known as 'regularisation', and the grid on which it is performed is called the *regularisation mesh* (de Groot-Hedlin and Constable, 1990). For example, in this case, we can write the *roughness function*, $Q(y_i)$, as a scaled norm of the Laplacian beneath site y_i (Smith and Booker, 1991):

$$Q(y_i) = \int \left(\frac{d^2 m(y_i, z)}{df(z)^2} + g(z) \frac{d^2 m(y_i, z)}{dy^2} \Big|_{y=y_i} \frac{d^2 z}{df(z)^2} \right)^2 df(z), \qquad (7.6)$$

where $g(z)$ allows for a trade-off between penalising horizontal and vertical structures, and $f(z)$ controls the scale length for measuring structure at different depths. Choosing $m = \ln(\sigma)$ and $f = \ln(z + z_0)$ generally produces misfits between modelled and measured data that are uniform across the frequency spectrum (Smith and Booker, 1988). The non-dimensional weight function $g(z)$ can be generalised as (Smith and Booker, 1991):

$$g(z) = \alpha \left(\frac{\Delta i}{z + z_0} \right)^\eta, \qquad (7.7)$$

where Δi is the distance between measurement sites neighbouring y_i, and α and η are constants (that are unrelated to phase-sensitive strike and skew). As $\Delta i \to \infty$, $g(z) \to \infty$. Therefore, horizontal gradients are penalised more relative to vertical gradients the greater the site spacing. The constant α scales the horizontal versus vertical derivatives in Equation (7.7). The larger the value of α, the greater the degree of horizontal smoothing. Small values of η favour shallow structure, but as η is increased from 0 to 1.5, the penalty for horizontal structure becomes independent of depth. Default values of $\alpha = 4$ and $\eta = 1.5$ are suggested by Smith and Booker (1991).

The roughness operator and misfit are jointly minimised via the operator W_i:

$$W_i(y_i) = Q(y_i) + \beta_i e_i^2, \qquad (7.8)$$

where β_i is a trade-off parameter between model structure and misfit, and $\sum e_i^2$ is the standard χ^2 statistic. Since the inverse problem is non-linear, χ^2 is reduced in small steps over a number of iterations so that assumptions inherent in the linearisation (e.g., Equation (7.2)) are not violated.

Changes in the model structure that are required by data at (an)other site(s) can produce a local increase in W_i. This is particularly true for **B**-polarisation data, which are sensitive to electric currents flowing along the modelled profile. A global measure, W_G, of the misfit function expressed in Equation (7.8) that is not dominated by sites with very large or very small β_i is (Smith and Booker, 1991):

$$W_G = \sum_{i=1}^{isites} \frac{W_i}{\beta_i + \beta_{median}}. \tag{7.9}$$

Convergence to the model with the smallest roughness commensurate with an acceptable (minimum) misfit requires several iterations (typically 10–20). If a priori knowledge about the presence of sharp discontinuities in resistivity is available, then the penalty for roughness can be removed at cells that border on these expected discontinuities (e.g., de Groot-Hedlin and Constable, 1990).

A flowchart of the steps involved in the inversion process is shown in Figure 7.2. It is advisable to experiment with different starting models to ensure that a similar solution is reached independent of the starting model. Smith and Booker (1991) suggest producing a 2-D model by introducing data components over a series of RRI inversions (Figure 7.3). The computation time required for RRI is less than for the Occam inversion scheme, because in the former lateral electric and magnetic field gradients calculated from the previous iteration are incorporated into the forward-modelling stage as an approximation to the new fields, and the forward model responses are computed as successive perturbations on a string of 1-D models that represent the structure beneath individual sites. Another popular 2-D inversion scheme, which is simple to install and implement has been programmed by Mackie *et al.* (1997).

Some inversion algorithms (e.g., Smith and Booker, 1991; de Groot-Hedlin, 1991) allow *static shift* parameters to be calculated simultaneously with the modelling procedure by favouring incorporation of surface structure at adjacent sites that minimises horizontal conductivity gradients at greater depths in the model (see Section 5.8).

7.3 Artefacts of inversion

Let's consider what happens if nominally 2-D, but anisotropic, model data are inverted to reproduce a 2-D model. Figure 7.4(c) shows a least-structure inverse model of data sampled along a profile of twelve equidistant, synthetic MT sites, which were synthesised from a 2-D forward model (Figure 7.4(a)) containing an

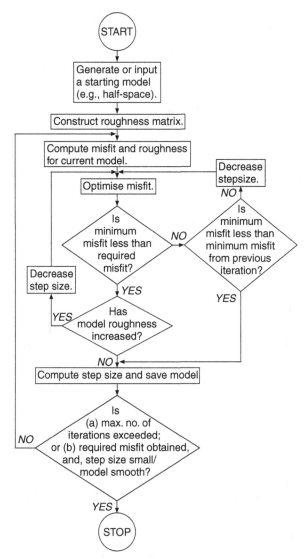

anisotropic layer between 12 and 22 km depth, embedded within
a layered half-space. The synthetic data containing anisotropy
(Figure 7.4(b)) were generated using the 2-D forward model code
from Wannamaker *et al.* (1986), and were then inverted using a 2-D
inversion routine from Mackie *et al.* (1997). The data (Figure 7.4(b))
from all twelve hypothetical sites are equivalent, because they all
image exactly the same 1-D layered structure with an anisotropic
layer composed of a regular sequence of alternating dykes having
resistivities of $5\,\Omega\,m$ and $50\,\Omega\,m$ at 12–22 km depth (Figure 7.4(a)).

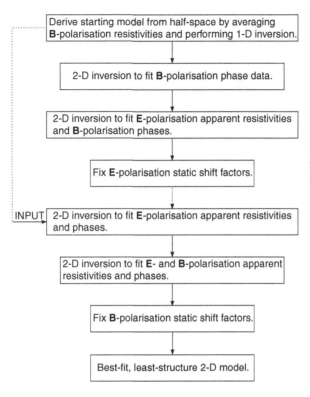

Figure 7.3 Flowchart showing successive introduction of **E**-polarisation and **B**-polarisation apparent resistivities and impedance phases into the inversion process, following a scheme suggested by Smith and Booker (1991) for implementing RRI inversion, which incorporates a weighting criterion for estimating static-shift factors.

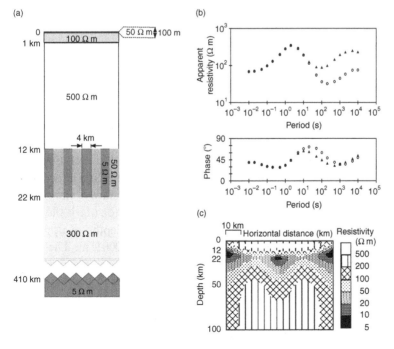

Figure 7.4 (a) Layered model containing 10-km-thick (12–22 km) anisotropic layer with infinite lateral extension. (b) Apparent resistivities and impedance phases generated by 2-D forward modelling of (a). (c) 2-D least-structure model obtained by inversion of twelve evenly spaced synthetic data sites, each having the anisotropic apparent resistivities and impedance phases shown in (b).

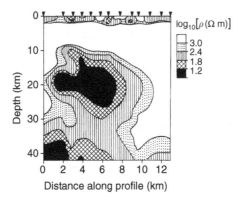

Figure 7.5 2-D least-structure model generated by inverting data measured along a profile approximately perpendicular to the geological strike in NE England (see Figure 5.12(b)). Arrows along top of model indicate site locations.

The individual dykes composing the anisotropic layer cannot be resolved since their respective lateral dimensions are much less than the depth to their tops.

The 2-D inversion (Figure 7.4(c)) yields high-conductivity blobs, with resistivities of between 5 and $10\,\Omega\,\text{m}$, within larger blobs of between 10 and $20\,\Omega\,\text{m}$ within still larger blobs of between 20 and $50\,\Omega\,\text{m}$. These blobs arise owing to the difficulty associated with fitting both polarisations of the apparent resistivities and impedance phases jointly.

Below the profile itself, we appear to have a basin-like structure that dips at an angle of $\sim\!20°$ on either side, but we know that this again is purely an artefact of the inversion. There was no dipping structure in the original model.

The scenario in Figure 7.4 is one of the simplest examples of why we have to be careful about taking the results of 2-D inversion too literally. In this example, the original model had a fixed 2-D *electromagnetic strike*, so that **E**- and **B**-*polarisations* can be decoupled. This is often not the case with measured data. Figure 7.5 shows a 2-D model generated by inverting data measured along a profile approximately perpendicular to the geological strike in NE England (Figure 5.12(b)). The modelled apparent resistivities and impedance phases of the two polarisations lie within two standard deviations of the measured data. The model shows a conductor (dark blob) of resistivity $10\,\Omega\,\text{m}$ extending downwards from depths of approximately 12 km, in a background resistivity of $1000\,\Omega\,\text{m}$. The dark blob has horizontal dimensions of 5–6 km, which at a depth of 10–12 km is below the theoretical resolution limit of the data. The blob is an artefact that arises only because the inversion scheme applied is required to fit divergent apparent resistivities and impedance phases (Figure 7.6) via a 2-D model. The divergence of the

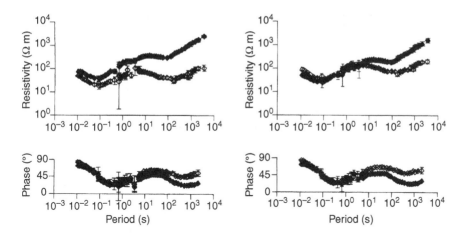

Figure 7.6 Apparent resistivities and impedance phases for two MT sites along the profile from which Figure 7.5 is derived. (Redrawn from Simpson and Warner, 1998.)

differently polarised apparent resistivities and impedance phases probably arises owing to the imaging of the edge of a resistive granitic upper-crustal block, which bounds more conductive sedimentary basins, and owing to the proximity of the coastline – just 45 km away to the east, and 90 km away to the west. Although relatively shallow (50–70 m), the North Sea is underlain by thick (possibly > 5 km deep) sedimentary grabens. By incorporating the onshore and offshore conductivity structures into a 3-D model, and making reasonable assumptions about the conductances of the seawater and sedimentary basins, the observed impedance phase splitting can be reproduced, without the need for a conductive blob in the mid crust. In fact, a mid-crustal conductance of less than 200 S (that's more than a factor of 10 less than in the 2-D blob model (Figure 7.5)) is sufficient to explain the data. Consideration of the lateral conductivity structures also explains the observation that the electromagnetic strike in NE England is period dependent (Figure 5.12(a)).

In our case study, using an average electromagnetic strike for the purpose of performing 2-D inversion of the MT data resulted in a model containing a conductive blob having horizontal dimensions that are not truly resolvable by the input data, and which on further investigation with 3-D forward modelling was found to be an artefact. Models containing blobs occur frequently in electromagnetic literature. The blobs apparent in these models arise primarily owing to an inability to fit both principal polarisations of the impedance tensor jointly using 2-D inversion. Anisotropy or three-dimensionality may give rise to this problem: be wary of models containing blobs at depth with lateral dimensions that are not resolvable by the input data!

7.4 Introducing a priori information

When only mathematical misfit criteria are considered, a range of models can always be found that fit the data equally well. By constraining the range of physically acceptable models that fit a given dataset, a priori information may help to overcome the non-uniqueness problem to some extent. Consider the model shown in Figure 2.5. Would unconstrained, least-structure inversion of data generated by such a model be appropriate? Certainly, not! For this type of scenario, it would be appropriate to introduce into our model scheme our a priori knowledge about the existence of a sharp contact.

A priori information may come from geological sources (as in the case study from NE England discussed in Section 7.3), or be inferred from other geophysical surveys. A priori information is easily incorporated into forward models. Most inversion routines also enable a priori information to be incorporated, by allowing user-defined conductivities to remain fixed at specified locations during inversion. However, it is important to avoid the temptation of constructing a model based on assumed geological structure, and then inferring from an adequate fit to the measured data that all of the features in the model are required by the data, rather than being merely allowed by the data. Therefore, it is generally better to combine the use of a priori data with *hypothesis testing*. For example, velocity contrasts in a seismic model may have been interpreted as arising owing to the presence of a few percent aqueous fluid. So, we might try incorporating an appropriate conductance into our MT model at the depth indicated by the seismic model, and ask whether we can fit our MT data better with or without the hypothesised conductance.

7.5 Forward modelling versus inversion

Ideally, inversion and forward modelling should be used in conjunction with each other. Inversion can be employed to produce initial approximations to the Earth's conductivity structure, but only through forward modelling do the limitations and strengths of MT data become clearly apparent. Unfortunately, many MT practitioners model data using inversion algorithms exclusively, and are content to compute one 'best-fit' model for their data. Also, because only 1-D and 2-D inversions are readily practicable at present, 3-D induction effects are often ignored.

Always be wary of the non-uniqueness problem inherent to modelling MT data. Do not restrict to a single inversion of MT data, but run multiple inversions. Any MT model derived via inversion can and should be explored further using forward modelling. How sensitive are the model data to perturbations in the modelled conductivity structure? What happens if a model with more focussed boundaries is substituted for a least-structure one? Could any of the modelled conductivity structures at depth arise from shallower lateral conductivity contrasts and three-dimensionality?

7.6 Avoiding common mistakes

Because electromagnetic energy propagates diffusively, magnetotelluric sounding resolves conductivity gradients rather than sharp boundaries or thin layers. For this reason, most 2-D inversion schemes are founded on the philosophy of minimising model complexity, wherein, rather than fitting the measured data as well as possible (which maximises the roughness of the model), the smoothest model that fits the data to an accepted tolerance threshold is sought. Whilst least-structure models reduce the temptation to over-interpret data, it should be remembered that sharp boundaries (e.g., the contact between an ore body and the resistive host rock into which it is intruded) do occur within the real Earth. Therefore, the models produced using least-structure inversion schemes might be viewed as depicting blurred, rather than focussed images of the real Earth structure. As such, these images should not be interpreted too literally.

Because MT data resolve conductors better than resistors, the bottom of a conductor cannot generally be resolved. Thus statements such as 'The conductor extends to a depth of . . .' may correctly describe the modelled image, but are misleading physically. Whereas, conductances can be well-constrained by MT data (Weidelt, 1985), resistivities and thicknesses of structures generally are not (see Figure 6.4 and Figure 6.5). For this reason, it is better to consider the conductance of conductive bodies, rather than to describe them in terms of the resistivities and thicknesses used to parameterise them in the model. Also, be aware how the resistivity scale chosen to visualise models can lead to subjective bias during their interpretation (see Section 4.4), and that the choice of resistivity contour used to delineate a feature will be open to a degree of arbitrariness.

Most 2-D inversions give an overall misfit of a model to an input dataset, as well as misfits for data at individual sites. Before interpreting any conductivity anomaly, it is always wise to consider where the model fits the data best and where the misfit is higher.

Also, beware the anomaly that is supported by only a few data points from an individual site!

The least-square misfit statistic expressed in Equation (6.6) depends on the errors in the measured data, making the utility of a model that has an acceptable fit to measured data reliant on the size of data errors, and the precision with which they have been estimated. Data exhibiting a high degree of scatter, but large errors, might be fitted quite easily with a 2-D model, even if strong 3-D effects are present, whereas 2-D data with errors that are underestimated may prove difficult to fit within a specified tolerance.

Always check the maximum and minimum *penetration depths* achieved by the data. All 2-D inversions assign resistivity values to cells embedded in a user-defined grid. Resistivity values will still be assigned to cells, even if no constraints exist! Some 2-D inversions also calculate sensitivities. These are useful parameters that help the user to verify how well particular regions of the model are constrained by the available measured data. In general, a conductor having conductance less than the overburden conductance will be ill resolved.

Never stop at computing a single model that fits the dataset, and always investigate 'preferred' models output by inversion algorithms further using forward modelling. A number of parameters (e.g., starting model, and α and η in Equation (7.7)) in the inversion can be modified, and the inversion re-run in order to investigate which structures are robust, in the sense that they retain their geometry, etc. It can also be a good idea to investigate how perturbations in the model grid influence the inversion/forward model, and to verify that the model is adequately gridded. Some general rules for generating adequate model grids can be found in Section 6.2.1.

In many 2-D inversions, only apparent resistivities and impedance phases for two principal polarisations are modelled. Don't neglect to consider how induction arrows, geomagnetic depth sounding (GDS) transfer functions, and diagonal elements of the impedance tensor might fit into the model.

Whilst some 2-D inversion schemes can synthesise an anisotropic environment, others do not allow for this possibility directly. One method that has been adopted to model an anisotropic environment is to invert data from the two principal polarisations independently, thereby producing two models with different conductivities in particular depth ranges. This approach is the easiest way to fit MT data for which the apparent resistivities and impedance phases of the two principal polarisations differ. However, there should be justification for preferring anisotropy at depth to shallower lateral conductivity contrasts, other than ease of modelling. On the other hand, as shown in Section 7.3, if anisotropy is ignored, and both principal polarisations of the data are inverted simultaneously, the ensuing model may contain blobs that are merely diagnostic of anisotropy.

Because of their wide availability and ease of implementation, 2-D inversion schemes are a favoured tool of MT practitioners. In contrast, although 3-D forward modelling is computationally viable, obvious 3-D effects in measured data are often ignored owing to the lack of a viable 3-D inversion scheme. Be prepared to ask yourself, 'To what extent is 2-D inversion appropriate for my dataset?'

Modelling of measured MT data is always non-unique. Therefore models are most useful when they are used to test hypotheses or to challenge existing formulations, and least useful when they are interpreted too literally.

Chapter 8
The general link to other geosciences: conduction mechanisms

Laboratory measurements of the electrical conductivity of mineral assemblages thought to comprise the transition zone in the mid mantle have been performed. The results agree very well with conductivity models of the transition zone derived from magnetotelluric data. This is surprising given that laboratory measurements are performed on small samples, whereas magnetotelluric sounding samples over a huge volume, and implies that the mid mantle is reasonably homogeneous with respect to conductivity. The mantle is a semiconductor and its conductivity generally increases with depth, as does its temperature.

Magnetotelluric sounding curves often indicate a zone of high conductivity in the mid- to lower-continental crust. This high-conductivity layer might, for example, be modelled as a 5-km-thick, $0.1\,S\,m^{-1}$ layer. However, there is no Earth material with a conductivity of $0.1\,S\,m^{-1}$, except brine, and we don't expect a 5-km-thick liquid layer in the crust. Rather, the modelled conductance represents a mixture comprised of a resistive rock matrix, and a more conductive second component. There is an ongoing debate as to the nature of this second component, which may promote electrolytic conduction in a network of cracks filled with concentrated brines, or electronic conduction in a network of graphite or ores. Mixing laws, scaling techniques and models of crack connectivity lead us to a more quantitative understanding of the conduction mechanism. It seems inappropriate to pose the question 'fluids or graphite?' in a global sense, but in some cases, the amount of conductance, and possible anisotropy thereof, may help to distinguish between fluids or graphite as principal causes of conductivity (though both fluids and graphite may still be present).

There are first hints towards regional conductivity anomalies in the uppermost mantle, below the lithosphere. These anomalous regions of conductivity cannot be explained by semi-conduction in dry olivine. Therefore, water, melt or carbon may be present in the sub-lithospheric mantle.

8.1 Laboratory measurements under *in-situ* conditions

Laboratory studies indicate that the electrical properties of Earth materials are strongly dependent on frequency, temperature and point-defect chemistry. The conductivities of Earth materials can be determined using impedance spectroscopy (e.g., Roberts and Tyburczy, 1991). Impedance is measured as the total opposition to current flow in response to an AC signal. One experimental setup involves monitoring the voltage drops across a circuit containing a sample and a high-precision resistor, through which an AC signal is passed, using high-input impedance, low-noise amplifiers. The imped-ance of the sample at a given frequency of AC signal is then determined from the measured potential difference and current, using matrix inversion. In rocks, impedance normally consists of both resistive and capacitive components, making it a complex quantity. Real and imaginary parts of the impedance form imped-ance arcs (semi-circular arcs approximately centred on the real axis) when plotted in the complex plane (Figure 8.1). Each imped-ance arc corresponds to one or more separate *conduction* processes that has a different characteristic relaxation time ($\tau = R \times C$, where R is DC resistance and C is capacitance), and can be modelled using circuits of resistors and capacitors. The basic electrical circuit for producing an impedance arc consists of a resistor in parallel with a capacitor.

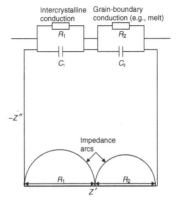

Figure 8.1 The basic principles underlying impedance spectroscopy. Real, Z', and imaginary, Z'', components of the impedance of a rock sample, when plotted in the complex plane, produce semi-circular traces called impedance arcs, which can be modelled using circuits composed of resistors and capacitors.

Figure 1.5 summarises the electrical conductivities of some common Earth materials. MT studies around the world indicate that the deep crust and mantle are often more conductive and more electrically anisotropic than anticipated from laboratory measurements on uplifted lower-crustal rocks (e.g., Kariya and Shankland, 1983) or dry olivine (Constable *et al.*, 1992), respectively. For example, the anisotropy of electrical conductivity of mylonites from the deep crust (Siegesmund *et al.*, 1991) is too low to explain most observations of deep-crustal electrical anisotropy (Figure 8.2). *In situ* crustal conductivities may be enhanced by saline fluids and/or graphite, metallic oxides and sulphides, or partial melt. Mantle conductivities may be enhanced by fluids (including partial melts), graphite or hydrogen diffusivity (see Sections 9.2 and 9.3). Macroscopic anisotropy is most easily explained if conductive phases are preferentially aligned (e.g., owing to crystal-preferred orientation) or exhibit a higher degree of interconnection in the more conductive direction.

Graphitic films have been observed in metamorphic rocks using Auger electron emission spectroscopy (Frost *et al.*, 1989). Amphibolite and gneiss core samples from depths of between

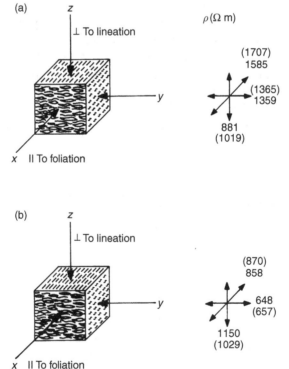

Figure 8.2 Macroscopic fabric elements of mylonite rock samples from the Indian Ocean ridge and their direction-dependent electrical resistivities. The values in brackets are for low pressures, corresponding to shallow crustal depths, whilst the other values are for the high pressures expected at deeper crustal depths. The resistivities of these rock samples are too high, and their anisotropies too low to explain deep-crustal conductivity anomalies revealed by MT field measurements. (Redrawn from Siegesmund *et al.*, 1991.)

1.9–7 km in the KTB deep borehole that were saturated with NaCl
solution exhibited increased conductivity in a direction at an angle
to foliation when pressurised in the laboratory (Figure 8.3), despite
the evacuation of fluid under pressure (Duba *et al.*, 1994; Shankland
et al., 1997). The anisotropic conductivity increase, was explained in
terms of reconnection of graphite under pressure. Ross and Bustin
(1990) demonstrated that shear stress promotes graphitisation of
rocks, whilst Roberts *et al.* (1999) found evidence of deposition of
carbon films on fracture surfaces during dilitancy of rocks under
controlled laboratory conditions. The phase transition from conducting graphite to insulating diamond occurs at temperatures of
~900–1300 °C and pressures of 45–60 kbar. Therefore, graphite is
unlikely to be the cause of conductivity anomalies deeper than
~150–200 km in the upper mantle.

Stesky and Brace (1973) demonstrated that serpentinised rocks
from the Indian Ocean ridge are 3–4 orders of magnitude more
conductive than serpentinite-free peridotite, gabbro and basalt

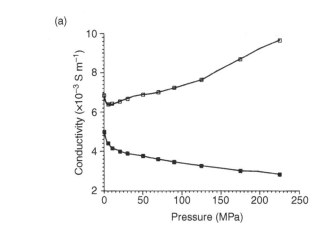

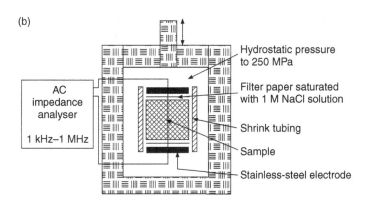

Figure 8.3 (a) Electrical
conductivity of an amphibolite
sample from a depth of
4.149 km in the KTB borehole.
Open symbols are
measurements at 30° to the
normal of the foliation plane
and closed symbols are for
measurements in the plane of
foliation. The conductivity in
the plane of foliation decreases
with increasing pressure,
whereas the conductivity in the
direction at an angle to the
foliation increases with
pressure, after decreasing with
pressure in the first ~10MPa.
The conduction mechanism
is interpreted to be the sum
of highly conductive
intergranular phases (such as
graphite) and saline fluids.
(b) Schematic diagram of
experimental setup used
to measure electrical
conductivities shown in (a).
(Redrawn from Duba *et al.*,
1994.)

from the same area. The enhanced conductivity may be associated with magnetite formed during serpentinisation. In addition to a conductivity anomaly, serpentinite should give rise to a magnetic anomaly.

Boreholes drilled to 12.3 km depth on Russia's Kola peninsula and to 8.9 km in Bavaria, southern Germany revealed that the rock at these mid-crustal depths was saturated with highly saline fluids – a most unexpected result. At mid-crustal temperatures and pressures, highly saline fluids can be expected to have conductivities of the order 25–50 S (Nesbitt, 1993).

Although mantle minerals are nominally anhydrous, they are able to incorporate dissociated H^+ and OH^- ions (Mackwell and Kohlstedt, 1990; Bell and Rossman, 1992). Measured concentrations of hydrogen content in natural samples indicate that pyroxene minerals are the most important reservoirs for hydrogen in the upper mantle, followed by garnet and olivine (Bell and Rossman, 1992). Hydrogen diffusion in the mantle could enhance the electrical conductivity of the mantle according to the Nernst–Einstein equation (Section 9.3, Equation (9.1)). Laboratory measurements indicate that hydrogen diffusivities vary depending on mineral type, and that they may be strongly anisotropic in single crystals.

Tyburczy and Waff (1983) showed that partial-melt conductivities are far less dependent on pressure than on temperature. Laboratory conductivity measurements on partially molten samples and mixing-law calculations are compared in Roberts and Tyburczy (1999). A compilation of results of laboratory electrical conductivity measurements on dry olivines and basaltic melts at a range of temperatures is shown in Figure 1.4. At upper-mantle temperatures, partial melt is approximately 1–2 orders of magnitude more conductive than dry olivine. However, many conductivity anomalies occur in stable regions at depths where the temperature is expected to be less than the solidus temperature of mantle silicates, and would require unrealistic quantities of partial melt to explain them. Based on a comparison of laboratory measurements made under controlled oxygen fugacities and electrical conductivity models of the Western Cordillera in northern Chile derived from MT data (Echternacht et al., 1997), at least 14% partial melt would be required below the magmatic arc in the central Andes to explain the observed deep-crustal conductance, which exceeds 40 000 S (Schilling et al., 1997).

Upper-mantle temperatures and pressures can be simulated in a multi-anvil press, in which 1–20 mm^3 samples are sealed from interaction with the external atmosphere, pressurised, and heated

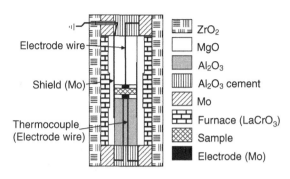

Electrode wire

Shield (Mo)

Thermocouple
(Electrode wire)

▤ ZrO₂
□ MgO
▓ Al₂O₃
▥ Al₂O₃ cement
▧ Mo
▦ Furnace (LaCrO₃)
▨ Sample
■ Electrode (Mo)

Figure 8.4 Experimental setup for complex impedance measurements under mantle conditions of pressure, temperature and oxygen fugacity. (Redrawn from Xu *et al.*, 1998.)

with a laser beam. A typical experimental setup for performing mantle conductivity measurements is shown in Figure 8.4. The molybdenum shield and electrodes help to minimise leakage currents and temperature gradients, and to control oxygen fugacity (fO_2). The importance of controlling fO_2 when determining electrical conductivities in the laboratory has been discussed by Duba and Nicholls (1973), Duba *et al.* (1974), Duba and von der Gönna (1994) and Duba *et al.* (1997).

The upper mantle is composed predominantly of olivine and pyroxene. Laboratory measurements indicate an ∼2 orders of magnitude increase in electrical conductivity arising from a phase change from olivine to wadsleyite, which is expected to occur at depths of ∼410 km (Figure 8.5; Xu *et al.*, 1998). Less dramatic increases in conductivity are expected at the wadsleyite to ringwoodite to prevoskite+magnesiowüstite transitions that are expected to occur a depths of ∼520 km and ∼660 km respectively. At upper-mantle temperatures and pressures olivine and pyroxene have similar electrical conductivities, whereas in the transition zone, the electrical conductivities of pyroxene and ilmenite–garnet compositional systems are 1–2 orders of magnitude lower than those of wadsleyite and ringwoodite (Xu and Shankland, 1999).

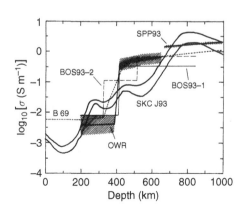

Figure 8.5 Conductivity distribution in and around the transition zone between the upper and lower mantle. Laboratory data: OWR (Xu *et al.*, 1998) and SPP93 (Shankland *et al.*, 1993). Magnetotelluric and geomagnetic models: B69 (Banks, 1969), BOS93-1, -2 (Bahr *et al.*, 1993) and SKCJ93 (Schultz *et al.*, 1993). (Redrawn from Xu *et al.*, (1998.)

8.2 How field measurements and laboratory measurements can or cannot be combined.

The electrical responses of Earth materials are frequency dependent (Roberts and Tyburczy, 1991). In polycrystalline materials, conduction occurs exclusively through grain interiors at frequencies higher than $\sim$100 Hz, whereas grain-boundary conduction also influences conductivities at typical MT sounding frequencies. Indeed, the DC conductivity of olivine is dominated by grain-boundary conduction at temperatures less than 1000 °C. Therefore, it is inadvisable to use laboratory measurements made in the kHz range, which may not yield results comparable to DC measurements, as a benchmark for interpreting long-period MT field measurements.

In order for conductive materials (e.g., graphite, saline fluid or partial melt) to enhance conductivities, the conductive component must form an interconnected network within the resistive host medium. Therefore, assumptions concerning the distribution of fluids within a host rock must be made before the quantity of saline fluids/graphite/partial melt required to generate a particular electrical conductance can be calculated. Archie (1942), considering the electrical resistivity of an electrolyte-bearing rock, proposed the empirical formula:

$$\sigma = \sigma_f \eta^n, \tag{8.1}$$

where η is *porosity*, σ_f is the conductivity of the electrolyte (e.g., brine) and n is an empirically determined exponent, varying according to the shape and interconnectivity of the pores. The geometric aspect embodied in n can be seen as analogous to the aspect-ratio dependence of seismic velocity–porosity models. Higher and lower degrees of pore interconnection have been characterised by smaller and larger Archie's Law exponents: 1 for interconnected, low-aspect-ratio cracks; 2 for poorly connected, high-aspect-ratio cracks (e.g., Evans *et al.*, 1982). Laboratory data measured under conditions approaching those thought to be representative of the mid crust suggest the likelihood of intermediate geometries, involving pore interconnection via grain-boundary tubes, and characterised by an Archie's Law exponent of 1.5 (e.g., Lee *et al.*, 1983). For very low porosities, it becomes necessary to take into account the conductivity, σ_r, of the rock matrix (Hermance, 1979). This is achieved by modifying Equation (8.1) to:

$$\sigma = \sigma_r + (\sigma_f - \sigma_r)\eta^2. \tag{8.2}$$

The interdependence of mixing ratios, connectivity and conductivity in multi-phase media will be discussed in more detail in Section 8.3.

At deep-crustal and mantle temperatures, minerals behave as semi-conductors, and their temperature-dependent conductivities can be described by an Arrhenius equation:

$$\sigma = \sigma_0 e^{-E_A/k_B T} \tag{8.3}$$

where E_A is the activation energy, k_B is the Boltzmann constant and σ_0 is the conductivity as temperature, $T \to \infty$.

The conductivities of saline fluids and partial melts are also temperature dependent (Figures 1.3 and 1.4). Hence, assumptions must be made concerning the likely temperature at the depth where a conductor is detected. Steady-state temperature–depth profiles calculated assuming 1-D heat conduction in the lithosphere are shown in Figure 8.6. Tectonic processes will modify these geotherms (e.g., Chapman and Furlong, 1992), which therefore describe temperature–depth distributions within the Earth only very approximately. Measurements made in deep-crustal boreholes on the Kola Peninsula and in southern Germany (KTB) indicated significantly higher temperatures than predicted. For example, the temperature at 10 km depth within the Kola borehole was found to be 180 °C, instead of a predicted 100 °C (Kozlovsky, 1984).

Laboratory measurements have shown that strain energy can promote graphitisation: shear stress facilitates graphitisation in the 400–500 °C temperature range (Ross and Bustin, 1990; Large *et al.*,

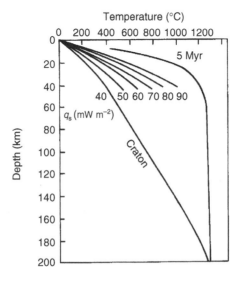

Figure 8.6 Framework geotherms for continental crust with different surface heat flows, q_s, calculated assuming steady-state conductive heat transfer (Chapman and Furlong, 1992), and approximate bounding geotherms for young (5 Myr) oceanic mantle and cratonic mantle (Davies and Richards, 1992).

1994), whereas graphite forms at temperatures of 600–2000 °C in the absence of shear stress. By providing an explanation for how graphite forms under particular deformation conditions, laboratory measurements have provided the missing link between MT studies, which often indicate conductivities within shear zones that are implausibly high to be generated by saline fluids, and the contention that high conductivities within the Earth might be caused by interconnected graphite.

Roberts *et al.* (1999) report the formation of graphite along new mineral surfaces during brittle fracturing in the presence of carbon gases at temperatures above 400 °C. However, they acknowledge that the strain rates applied to laboratory specimens (10^{-6}–10^{-5} s^{-1}) were orders of magnitude greater than typical in nature (10^{-16}–10^{-10} s^{-1}), raising the question if and how such experimental data should be extrapolated to geological conditions. Reconciliation of the timescales on which laboratory measurements can be conducted with geological timescales is a general scaling problem pertinent to all laboratory experiments involving a dynamic element.

In general, a higher degree of electrical anisotropy is observed from MT measurements than is apparent in rock samples tested in the laboratory. For example, core samples from the KTB deep borehole in southern Germany exhibit electrical anisotropy ratios of less than 1:8 (Rauen and Laštovičková, 1995), whereas MT data indicate electrical anisotropy ratios of 1:30 within the crust in the vicinity of the borehole (Eisel and Haak, 1999). One explanation for the scale dependence of electrical anisotropy is that a greater range of fracture sizes is sampled by an MT volume sounding than can be sampled when studying intact laboratory samples, which contain only fractures of dimensions smaller than the sample dimensions. The scale dependence of electrical anisotropy between laboratory and field scales is considered in more detail in Section 8.4. Scale dependence of electrical properties is particularly prevalent in the continental crust, because of its high degree of heterogeneity.

Laboratory measurements of the electrical conductivity of mineral assemblages believed to constitute the transition zone at 410 km depth in the mid mantle, and long-period MT data both imply a marked reduction in resistivity within the transition zone. Whereas high-pressure laboratory measurements are performed on minute samples, MT soundings at periods sufficient to penetrate to 410 km sample huge Earth volumes, yet models derived from these two very different sources of data agree within an order of magnitude (Figure 8.5).

8.3 Multi-phase systems and mixing laws: interdependence of mixing ratios and connectivity

In some cases, the conductance of a conductive body may help to distinguish the physical cause of enhanced conductivity. For example, a model derived from MT data might indicate the presence of a 15-km-thick layer having a resistivity of $1\,\Omega$m between depths of 15–30 km. Having heat flow data for the region where our MT survey was performed, we might calculate (see Section 8.2; Figure 8.6) that the temperature at the depth of the conductor is 350–450 °C. We do not know of an individual Earth component that should have a resistivity of $1\,\Omega$ m at temperatures of 350–450 °C. At mid-crustal temperatures, saline fluids are expected to have resistivities of 0.02–$0.04\,\Omega$ m (Figure 1.3). Remembering the inherent non-uniqueness of our MT model, and particularly the fact that MT resolves conductance better than resistivity and thickness, we calculate that a 300–600 m layer of saline fluids could generate a conductance equivalent to that in the model. However, a 300–600 m layer composed purely of saline fluids would be incapable of supporting the overlying 15 km of the Earth's crust. We must, therefore, infer that the conductance in our model arises as a result of a mixture of conductive and resistive phases.

The physical properties of composite media are calculated using mixing laws. Although this list is not complete, we shall distinguish three classes of mixing laws with respect to their application to electrical conductivity: minimum-energy theorems, effective (or equivalent) media theory and percolation theory.

8.3.1 Minimum-energy theorems

The two different media or 'phases' of which the composite is formed are represented by different resistors arranged to form a parallel or a series circuit through which a current is allowed to flow (Madden, 1976). In seismology, the equivalents of parallel and series averages are referred to as the Voigt limit and the Reuss limit, and stress is the variable that is analogous to current. In seismology, the physical properties that determine the characteristics of the bulk medium differ by only a few percent (consider the variation of seismic velocities within the Earth), and the Voigt and Reuss limits are widely applicable. On the other hand, the electrical properties of Earth materials can differ by 12 orders of magnitude, e.g., graphite $(10^{-6}\,\Omega\text{m})$ through to rock matrix $(10^{6}\,\Omega\text{m})$. In this case, the parallel average yields an unreasonably high conductivity, whilst the series average yields a conductivity that is biased too low.

8.3.2 Effective or equivalent media theory

The effect of inserting a single inclusion (e.g., a crack) into a medium is calculated and used to approximate the mean perturbation generated by a heterogeneous distribution of inclusions. The underlying assumption is that an heterogeneous medium can be treated as an homogeneous medium with an effective (bulk) property (e.g., conductivity). The conductivity of a two-phase effective medium is bounded by limits referred to as the *Hashin–Shtrikman upper and lower bounds* (Hashin and Shtrikman, 1962):

$$\sigma_f + (1 - \phi)\left(\frac{1}{(\sigma_m - \sigma_f)} + \frac{\phi}{3\sigma_f}\right)^{-1} > \sigma_{eff} > \sigma_m + \phi\left(\frac{1}{(\sigma_f - \sigma_m)} + \frac{(1 - \phi)}{3\sigma_m}\right)^{-1}$$

(8.3)

where σ_{eff} is the *effective conductivity*, σ_m is the conductivity of the solid matrix, σ_f is the conductivity of the high-conductivity material, and ϕ is the volume fraction of the conductive material. The upper bound, HS$^+$, describes the case that the conductive material is perfectly interconnected, whilst the lower bound, HS$^-$, describes the case that the conductive material is confined within isolated pockets. Hashin–Shtrikman upper and lower bounds are plotted in Figure 8.7, and illustrate the crucial influence of the degree of connectivity on the *effective conductivity* of a two-phase medium. Waff (1974) showed that for the case that $\sigma_f \gg \sigma_m$, the Hashin–Shtrikman upper bound can be approximated as:

$$\sigma_{eff} = \frac{2}{3}(\phi\sigma_f).$$

(8.4)

Figure 8.7 Normalised effective conductivity, σ_{eff}/σ_f, as a function of volume fraction, ϕ, of a conductive phase having a conductivity, σ_f, one thousand times higher than the conductivity, σ_m, of the host matrix. HS$^+$ and HS$^-$ are Hashin–Shtrikman upper and lower bounds (Equation 8.3), respectively. Dashed line shows Waff's (1974) approximation (Equation 8.4).

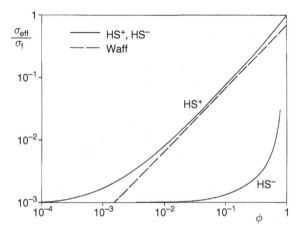

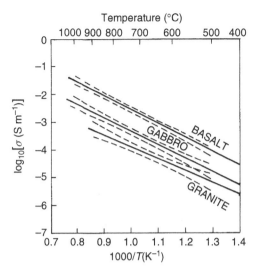

Figure 8.8 Best-fitting straight lines (solid) and their 95% confidence intervals (dashed) for laboratory measurements of the electrical conductivities of dry basalts, gabbros and granites as a function of temperature. (Redrawn from Kariya and Shankland, 1983.)

The condition that $\sigma_f \gg \sigma_m$ is valid for many realistic mixtures of Earth components, particularly those such as graphite or fluids distributed within a resistive rock matrix. Figure 8.8 presents laboratory measurements of three types of dry rocks as a function of temperature. From this graph we might, for example, deduce an appropriate rock matrix conductivity for mid-crustal rocks of $10^{-5}\,\mathrm{S\,m^{-1}}$. This is more than six orders of magnitude less than the conductivities expected for graphite or saline fluids at mid-crustal temperatures (Figure 1.5). On the other hand, mantle temperatures are higher, and the conductivity of dry olivine at mantle temperatures could play a role in the conduction mechanism, so that σ_m is no longer negligible in the mantle.

Effective media theory has been used with some success in seismic studies of porous media, in which elastic wavelengths are large compared to the dimensions of inclusions. However, in studying the electrical conductivity of two-phase media containing a dilute concentration of a conductive phase, the assumptions of effective media theory break down, because how a conductive phase is distributed (and how inclusions interact) generally has a greater influence on bulk conductivity than the volume fraction of conductive phase present. For example, a small amount of conductive phase can have a significant effect on bulk conductivity if it is perfectly interconnected, whereas a larger amount will have a negligible effect if it is contained in isolated pores. The breakdown of effective media theory is particularly pronounced in the vicinity of the *percolation threshold*.

8.3.3 Percolation theory

In cases of extreme heterogeneity, resistor networks can be used in order to explore the degree of connectivity and its impact on the effective conductivity. Further details of the application of percolation theory in MT studies are given in Section 8.5. Using a combination of effective medium theory and percolation theory, Waff's formula can be extended to allow for departures from the perfectly interconnected case, by introducing a dimensionless measure of electrical connectivity, γ (Bahr, 1997):

$$\sigma_{\text{eff}} = \frac{2}{3}(\phi\sigma_{\text{f}}\gamma) \quad 0 \le \gamma \le 1. \tag{8.5}$$

Waff (1974) considers the problem of connectivity in terms of liquid bridging. This is a statistical model in which the tortuosity of the path of the conductive phase can be visualised by means of a resistor network in which a fraction of resistors have been randomly (or pseudo-randomly) removed. The missing pathways represent the resistive rock matrix, whereas the resistors represent the conductive phase. The conductivity, σ, of a 2-D random resistor network of the sort represented by Figure 8.9 varies according to:

$$\left.\begin{array}{ll} \sigma = 2\sigma_0(P - 0.5) & 0.5 < P \le 1 \\ \sigma = 0 & P \le 0.5 \end{array}\right\}, \tag{8.6}$$

Figure 8.9 2-D resistor network with i and j cells in x- and y-directions, respectively.

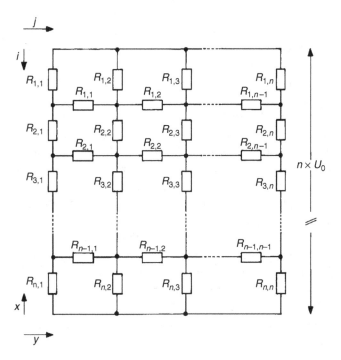

where σ_0 is the network conductivity for the case that all bridges are intact, and P is the fraction of remaining bridges (i.e., the percolation threshold, P_c, is 0.5 (see Figure 8.10)). For a cubic network, Equation (8.6) can be modified to (Shankland and Waff, 1977):

$$\left.\begin{array}{ll} \sigma = \dfrac{1}{0.69}\,\sigma_0(P-0.31) & 0.31 < P \le 1 \\[2mm] \sigma = 0 & P \le 0.31 \end{array}\right\}. \tag{8.7}$$

Equations (8.6) and (8.7) do not give accurate results in the vicinity of the percolation threshold (see, for example, Bahr (1997) and references therein).

8.4 Bulk conductivity versus direction-dependent conductivity

The MT technique yields conductances that are direction dependent (or anisotropic). Many early 1-D interpretations of MT data were based on *rotationally invariant* scalar averages of the *impedance tensor* that reduce direction-dependent conductances to a bulk conductance. Two examples of rotational invariants that have been proposed are the *Berdichevsky average*, Z_B (Berdichevsky and Dimitriev, 1976), which is the arithmetic mean of the principal (off-diagonal) components of the *impedance tensor*:

$$Z_B = \frac{Z_{xy} - Z_{yx}}{2}, \tag{8.8}$$

and the *determinant average*, Z_{det}, which is a geometric mean calculated from the square root of the determinant of the impedance tensor:

$$Z_{det} = \left(Z_{xx}Z_{yy} - Z_{xy}Z_{yx}\right)^{1/2}. \tag{8.9}$$

For example, based on a geometric average, the bulk conductance, τ_{bulk}, may be calculated as:

$$\tau_{bulk} = \left(\tau_{max}\tau_{min}\right)^{1/2}. \tag{8.10}$$

So far we have not emphasised modelling invariants, because although they were extensively used to obtain approximate average models of the Earth's conductivity structure when computational power was limited, their physical link to an Earth that we now recognise to be anisotropic and multi-dimensional is tenuous. With the advent of modern interpretation techniques it is no longer justifiable to reduce data to such a degree that important information about direction dependence of conductivity is masked.

Bahr (1997) has used fractal random networks of crack distributions to explain the scale dependence of anisotropy between laboratory and field scales. Small biases in the tendency for cracks to form in one direction rather than another can reproduce the observed scale dependence of electrical anisotropy between laboratory and field scales, suggesting that anisotropy might arise owing to a low degree of connectivity within a network that is close to the percolation threshold. From a rock mechanical viewpoint, a fracture network might be expected to remain close to the percolation threshold because fractures are propagated by fluid pressure, which drops dramatically as soon as the percolation threshold is reached (Guéguen *et al.*, 1991). For a conductive–resistive network to remain close to the percolation threshold, the presence of only small amounts of a conducting phase is implied. This model might, therefore, help to distinguish between the presence of significant quantities of fluids or minor quantities of graphite in some environments where high conductances are observed in one direction. For example, 5% by volume (5 vol.%) free, saline fluids would be required to explain the conductance of the conductive layer in the conductive-mid-crust model described in Section 3.1.1. Suppose this model were derived by inverting only one polarisation of the impedance tensor, and we were to realise subsequently that the model does not fit the other polarisation of the impedance tensor, which instead indicates a higher-resistivity layer in the deep crust. The reason for this dichotomy might be anisotropy. Would it be reasonable to suppose that 5 vol.% free, saline fluids could maintain interconnectivity in one direction, but not in the other?

Whilst the network models of Bahr (1997) provide one possible explanation for how electrical anisotropy can develop, they do not provide a physical explanation of the driving force that causes preferential formation of cracks in a particular direction. Maximum conductance often appears to be aligned perpendicular to present-day maximum stress, and sometimes perpendicular to a known palaeostress field. This observation may be related to shear-induced trapping of fluids or graphitisation along channels, or stress-induced anisotropic formation of cracks.

Simpson (2001a) developed a quantitative model in which conductance and anisotropy of conductance are jointly interpreted. To reflect the fact that MT data do not resolve conductivity and thickness, but rather conductance, Equation 8.5 may be rewritten as:

$$\tau_{\text{eff}} = \frac{2}{3}(\phi \, \sigma_{\text{f}} \, \gamma l) \qquad 0 \leq \gamma \leq 1, \qquad (8.11)$$

where τ_{eff} is the effective conductance and l is the thickness of the conductive layer. If only bulk conductance is known, then Equation (8.11) contains four unknown quantities (ϕ, σ_f, γ, and l), and only one known quantity (τ_{eff}). For the anisotropic case, we have:

$$\tau_{\max} = \frac{2}{3}\left(\phi_{\max}\sigma_{f,\,\max}\gamma_{\max}l_{\max}\right) \tag{8.12}$$

$$\tau_{\min} = \frac{2}{3}\left(\phi_{\min}\sigma_{f,\,\min}\gamma_{\min}l_{\min}\right). \tag{8.13}$$

Since $\tau_{\max}$ and $\tau_{\min}$ are attributed to the same layer (i.e., $l_{\max} = l_{\min}$) and the *porosity* within that layer is a bulk (volumetric) property (i.e., $\phi_{\max} = \phi_{\min}$), and the conducting phase is assumed to be the same for both conductances (i.e., $\sigma_{f,\,\max} = \sigma_{f,\,\min}$), Equations (8.12) and (8.13) can be re-written as:

$$\tau_{\min} = \frac{2}{3}\left(\phi_{\text{eff},\,\min}\sigma_f l\right) \tag{8.14}$$

$$\tau_{\max} = \frac{2}{3}\left(\phi_{\text{eff},\,\max}\sigma_f l\right) \tag{8.15}$$

where $\phi_{\text{eff},\,\max} = \phi\gamma_{\max}$ and $\phi_{\text{eff},\,\min} = \phi\gamma_{\min}$. The effective porosity, ϕ_{eff}, is the fraction of the porosity which contributes to conductance, and $\phi_{\text{eff}} \leq \phi$ since $0 < \gamma < 1$. The system of equations represented by Equations (8.14) and (8.15) is still underconstrained, because there are only two known quantities ($\tau_{\max}$ and $\tau_{\min}$) against four unknown quantities. However, by assuming (i) reasonable values for the conductivity of fluids ($\sigma_f = 35\,\text{S}\,\text{m}^{-1}$) and graphite ($\sigma_g = 3.5\times10^4\,\text{S}\,\text{m}^{-1}$), and (ii) that conductivity anomalies with their lids in the deep crust are generated within the crust and do not extend into the mantle (i.e., $l < (l_c - z)$ where l_c is maximum crustal thickness and z is the depth at which enhanced conductivity commences), we can investigate constraints on (i) the thickness of a conductive layer in the crust; (ii) the volume fraction of conductive phase required (Simpson, 2001a).

8.5 Scaling, random resistor networks and extreme heterogeneity

Resistor networks were first applied to MT data by Kemmerle (1977) in order to calculate the *distortion tensor*, $\underline{C}$, belonging to a 1-D impedance (Equation 5.19), using a priori knowledge of the distribution of surface conductance. Kemmerle (1977) obtained the spatial distribution of the near-surface conductance of the MT target area from dense DC measurements and represented it by a network of resistors.

A possible resistor network is shown in Figure 8.9. It consists of $n \times n$ resistors in the x-direction and $(n-1) \times (n-1)$ resistors in the y-direction. A voltage of $n \times U_0$ is applied in the x-direction. For a homogeneous network, in which all resistors have the same resistances, each of the resistors aligned in the x-direction is associated with a voltage U_0, whilst each of the resistors in the y-direction is associated with a zero voltage. For an inhomogeneous network, at each pair (i, j) of x, y resistors the distortion tensor elements:

$$c_{11} = U_{xi, j}/U_o$$
$$c_{12} = U_{yi, j}/U_o$$

(8.16)

can be ascertained if the voltage distribution within the network is known. After a rotation of the network model (or direction of the applied voltage) the two other elements c_{22} and c_{21} of the distortion tensor can be found. The voltage or current distribution in the network can be found by considering the following physical conditions: (1) U is a conservative potential in each of the $n \times (n-1)$ elementary cells; (2) The total current in the x-direction is the same in rows $1, \ldots, i, \ldots, n$ (this provides $(n-1)$ equations); (3) Kirchhoff's Law must be obeyed in the $(n-1) \times (n-1)$ nodes that provide linear independent equations; (4) the normalisation $U_x = n \times U_0$. Conditions (1)–(4) result in a total of $n \times n + (n-1) \times (n-1)$ equations, examples of which are given by Kemmerle (1977) and Bahr (1997). In a linear system:

$$\underline{\underline{R}}\mathbf{I} = \begin{pmatrix} 0 \\ 0 \\ 0 \\ 0 \\ U_o \end{pmatrix}$$

(8.17)

the current distribution in the $n \times n + (n-1) \times (n-1)$ resistors of the network is obtained by inverting the matrix $\underline{R}$. Once these currents are known, the voltages can be found by applying Ohm's law to every resistor.

Although originally designed to find the *static shift factors* in MT field data, the resistor network technique has rarely been used for that purpose. This is probably owing to the requirement that many DC or *TEM* measurements would have to be performed in the vicinity of every MT site. Where the method has been applied to ascertain static shift factors, comparison with the static shift factors obtained from long-period MT data (method (iii) from Section 5.8) has indicated discrepancies between the static shift factors obtained using the two correction techniques (e.g., Groom and Bahr, 1992). The most likely explanation for these discrepancies is that the underlying assumption that 'static shift is caused by a near-surface scatterer' is fallacious. Suppose that the static shift factors are estimated at periods of 20 000 s, which have *penetration depths* of approximately 300 km, using long-period MT data (method (iii), Section 5.8). Then, in order to act like a galvanic scatterer, a conductivity anomaly does not need to be at the surface – its depth should only be shallow compared to 300 km. If a conductive structure in the mid or lower crust (e.g. Section 9.1) is very heterogeneous, it can act like a galvanic scatterer on long-period data, but it cannot be detected with a DC or *TEM* method that only scans the near-surface.

The resistor networks used for estimating the static shift factors represent a map of the surface conductance, and therefore have a concrete geophysical meaning. Random resistor networks have also been used to explore the connectivity γ (introduced in Section 8.3) of the conductive component in two-phase systems. These random resistor networks are not envisaged as representations of geological units, but rather as any piece of Earth material. If resistors with low and high resistivities are used to represent the rock matrix and the conductive phase, respectively, then the connectivity of a resistor network is a non-linear function of the fraction, P, of low-resistance resistors. In terms of percolation theory, the resistor network in Figure 8.9 is regarded as an example of bond percolation (see, for example, Guéguen and Palciauskas, 1994) in a 2-D lattice. The connectivity, γ, in Figure 8.10 was calculated by means of a statistical experiment: P was varied in small steps, and for each P many realisations of the network were produced, in which the resistors were distributed over the nodes with a random generator.

Suppose the resistors representing the rock matrix are infinitely resistive or, for the purposes of numerical calculations, that they are many orders of magnitude more resistive than the conductive phase. If the fraction of resistors representing the conductive phase is below the percolation threshold, no current flows and the connectivity, which is a normalised (dimensionless) current flow through the network, is zero (Figure 8.10). In this model, the direction-dependent connectivity introduced in Section 8.4 stems from the extreme non-linearity of the connectivity (probability) function in the vicinity of the percolation threshold (Figure 8.10). Another finding is that the heterogeneity of the voltage distribution in the network is maximal in the vicinity of the percolation threshold.

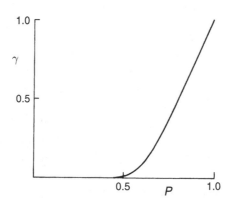

Figure 8.10 Electrical connectivity (normalised current), γ, as a function of the fraction, P, of low-resistance resistors in a 2-D resistor network.

Intuitively, this can be expected, given that there is no voltage heterogeneity in the trivial cases of homogenous networks ($P = 0$ or $P = 1$ in Figure 8.10). This area of research again touches upon the subject of distortion of MT data, since distortion imposed on long-period data by a heterogeneous conductor in the mid crust can also be described in terms of a voltage distribution – in this case, distributed over the sites of an MT array rather than at the nodes of a resistor network. Recently, an attempt has been made to compare these two voltage distributions and their heterogeneities, in order to test the hypothesis that the conductor in the mid crust forms an extremely heterogeneous network that stays close to the percolation threshold (Bahr *et al.*, 2002).

Heterogeneity with respect to electrical properties in two-phase 3-D random media has been studied by Kozlovskaja and Hjelt (2000), Hautot and Tarits (2002) and Bigalke (2003). Kozlovskaya and Hjelt (2000) imposed a fractal geometry on the grain-to-grain contact structure of aggregates and showed that both elastic and electric properties depend non-linearly on the degree of contact between grains. Hautot and Tarits (2002) generated synthetic two-phase models with random '*porosity*' and used the 3-D MT forward-modelling code of Mackie *et al.* (1993) to calculate electromagnetic data, and the DC (geoelectric) forward-modelling code of Spitzer (1995) to calculate electric data. They found that MT and DC data did not provide the same effective conductivity at a particular porosity; instead, they defined an effective interconnected volume fraction which is the same for EM and DC data. Bigalke (2003) generated synthetic EM, DC and *TEM* data for similar models and found that bulk conductivities calculated from the synthetic MT data are smaller then those calculated using the other two techniques.

Chapter 9
The special link to other geosciences

Is 'tectonic interpretation of conductivity anomalies' a useful concept? Probably yes, if over-simplifications are avoided. For example, there are cases where a high-conductivity layer in the mid crust also exhibits anomalously high reflectivity, and the combination electromagnetics and seismics has been advocated as a tool for 'detecting' water in the crust. However, application of the techniques described in Chapter 8 and petrological considerations suggest that, in old, stable regions, the lower crust is generally dry.

The combination of long-period MT and surface-wave seismology have revealed concomitant directions of seismic and electrical anisotropy, apparently supporting a hypothesis involving alignment of olivine crystals. How is such an alignment realised and maintained?

A collaborative project involving geodynamics, seismology and long-period MT sounding has been established in search for an intra-plate plume under Central Europe. The seismological part of the experiment suggests some evidence for a 150–200 K temperature increase in the uppermost 200 km within a 100 km by 100 km region. However, not only is the electrical conductivity within that volume only moderately increased, but also the strongest conductivity increase occurs outside the velocity anomaly. In addition, the electrical anomaly exhibits strong electrical anisotropy. Is this another hint that if melt is present, then this melt is not the dominant conductor, and that the bulk conductivity of the mantle is more strongly influenced by another conductive phase?

Few ocean-bottom MT studies have been performed and investigations into possible differences between the physical compositions of oceanic and continental mantle have begun only recently. Studies involve simultaneous measurements on a continent and on the ocean floor or, alternatively, on islands. Is the oceanic mantle more conductive? There is no definitive answer – a great encouragement for those who are interested in continuing field surveys of this type.

MT sounding can also be used to map faults. Since, (for reasons discussed in Chapter 8) both active and passive faults can be conductive, we have no way of distinguishing between active (with a potential to induce an earthquake) and passive faults on the basis of their conductivity. However, the orientation of a major fault zone with respect to the direction of the stress field acting on it is likely to determine whether a fault slips, and to some extent how far, as well as its potential sense of motion. Therefore, it could be useful to identify faults and their directions, prior to having the focal plane solution calculated after an earthquake. Then, although we can't use electromagnetic techniques to predict that an earthquake is going to occur, we might be able to work with structural geologists and civil engineers to predict the likely effects of an earthquake if slip occurs. Insurance companies use information of this type, when they calculate insurance premiums in earthquake regions.

Magnetotelluric sounding has also been employed in petroleum-bearing and geothermal regions, and has proven particularly useful where surficial salt domes, volcanics or karsts hamper the resolution of seismic techniques owing to absorption of sonic energy and reverberations in such geological complexes.

9.1 EM and seismics

In addition to generating conductivity anomalies, some Earth materials, such as aqueous fluids and partial melt, refract or reflect sonic energy, so that their presence can also be detected from variations in seismic characteristics (e.g., seismic velocity). A number of studies involving coincident MT and seismic refraction/reflection sites around the world have demonstrated that joint interpretation of electromagnetic and seismic data can provide tighter constraints on the physical causes of anomalies than can be gained from a single technique. Where both electrical conductivity and seismic velocity anomalies can be imaged, this provides independent information for quantifying the amount and nature of a physical phase responsible for generating both the electrical conductance and seismic characteristics. Lack of a seismic anomaly where an electrical conductivity anomaly exists, or vice versa, also has implications for physical interpretation.

During the 1980s, seismic reflection studies revealed laminations in the lower continental crust generated by an alternating pattern of high- and low-velocity layers, in many regions of the world (e.g., Allmendinger *et al.*, 1983; Mathur, 1983; BIRPS and ECORS, 1986; DEKORP, 1991). Correlations between zones of lower-crustal layering and enhanced electrical conductivity promoted speculation that both the seismic and electrical conductivity anomalies are generated by saline fluids (e.g., Hyndman and Shearer, 1989; Merzer and Klemperer, 1992). However, some of the correlations are contentious, because the seismic and magnetotelluric profiles that were correlated were rarely coincident, and because the possibility of *static shift* in MT data can lead to poor constraints on the depth to the top of a conductive layer, whilst the bottom of a conductive layer is inherently poorly constrained. Simpson and Warner (1998) challenged the idea of concomitant zones of seismic layering and electrical conductivity arising from saline fluids by evaluating the conductance of the lower crust below MT profiles lying coincident with P- and S-wave seismic reflection profiles in Weardale, NE England (see Figure 5.12 (b)). The low conductance of the lower crust implied the presence of less than 1% aqueous fluid, whereas the v_P/v_S ratios would require $\sim$6% fluid, implying that fluids are not the cause of seismic layering in Weardale. The laminations may arise from magmatic underplating (Furlong and Fountain, 1986). Petrological studies (Yardley, 1986; Yardley and Valley, 1997) also suggest that the lower crust in stable tectonic regions, such as Weardale, is likely to be dry. In contrast, extensive reservoirs of fluids are likely to be present in the deep crust in active orogens. A zone of bright seismic reflectors at depths of 15–20 km below the Tibetan plateau, which is associated with a conductance of 6000 S extending from the same depth range, has been ascribed to aqueous fluids and partial melt (Li *et al.*, 2003). Highly reflective zones with elevated v_P/v_S below the Bolivian Antiplano also correlate well with the upper boundary of an electrical conductivity anomaly, and both geophysical signatures have been interpreted in terms of substantial partial melting (Brasse et al., 2002). Spatially correlated zones of low P-wave velocity and high electrical conductivity below Tien Shan, Kirgistan have been attributed to aqueous fluid released by dehydration reactions in an amphibolite-rich mid crust that was thermally reactivated during the Cenozoic (Vanyan and Gliko, 1999).

9.2 EM and seismology

Application of EM and seismology to the same target area and depth has so far been restricted to a few case studies. However, where EM and seismological measurements have been performed jointly, the two geophysical techniques have proved complementary. At certain depths, (e.g., at the base of the lithosphere or at the 410 km discontinuity) compositional or physical changes in mantle materials might be expected to result in abrupt changes of both seismic velocities and electrical conductivity. Changes associated with the lithosphere–asthenosphere boundary or the mid-mantle *transition zones* generally result in increased conductivity with depth, and the problem of finding the depth to the transition is identical to determining the upper boundary of a conductive structure – a problem for which passive EM techniques are well suited (Section 2.5). The asthenosphere has been identified as a zone of low seismic velocity and enhanced electrical conductivity (e.g., Praus *et al.* 1990; Jones, 1999). Early attempts to jointly interpret these asthenospheric signatures rested on the idea that a few percent partial melt, which influences elastic, anelastic and electrical properties of rocks (e.g., Schmeling, 1985; 1986), existed in the asthenosphere. However, we now believe that it is unlikely that the continental geotherm crosses the dry mantle solidus at the depth in which the asthenosphere is imaged in many regions. The idea that partial melt is primarily responsible for the conductivity increase at the base of the lithosphere is also challenged by quantitative conductance models, because reasonable amounts of partial melt cannot explain the high conductance found in some field studies (e.g., Leibecker *et al.*, 2002; see also Section 9.3). The physical properties of the asthenosphere may rather be explained by the presence of small amounts of water. Hydrolytic weakening can readily explain the low viscosity of the asthenosphere (Karato and Jung, 2003), and hydrogen diffusivity can significantly enhance electrical conductivites (Karato, 1990). In regions where the mantle geotherm crosses the wet mantle solidus, partial melting may actually increase the viscosity and reduce the electrical resistivity of the asthenosphere owing to dewatering of the sub-lithospheric mantle.

In olivine, which is the dominant mineral in the upper mantle, hydrogen diffusivities are anisotropic (Kohlstedt and Mackwell, 1998), and alignment of olivine owing to viscous drag at the base of the lithosphere may lead to electrical anisotropy (Lizzaralde *et al.*, 1995). Electrical anisotropy at the base of the electrical lithosphere has been detected from MT measurements in central Australia (Simpson, 2001b; see Section 9.3). The electrical anisotropy below central Australia occurs at the same depth ($\sim$150 km) as SV-wave anisotropy, and the direction of high conductivity matches the fast direction of SV-waves (Debayle and Kennett, 2000). However, the seismic high-velocity lid extends deeper than the depth to the anisotropic layer (Gaherty and Jordan, 1995).

Kurtz *et al.* (1993) and Mareschal *et al.* (1995) also detected evidence of electrical anisotropy in Archaean mantle where seismic anisotropy has been detected (Sénéchal *et al.*, 1996). In this case, the direction of maximum conductivity and the polarisation direction of the fast shear wave are oblique to each other, and the electrical anisotropy has been attributed to shape-preferred orientation of graphite within ductile shear zones in the lithospheric mantle (Ji *et al.*, 1996).

EM can provide an important tool for resolving and constraining mantle anisotropy for three reasons.

1 The *electromagnetic strike* estimation can be very accurate and can result in *confidence intervals* as small as 7° (Simpson, 2001b).
2 Comparing the maximum and minimum conductances allows the strength of the anisotropy to be quantified.
3 Provided that the static shift problem (Sections 5.1 and 5.8) can be addressed, the depth of the anisotropic structure can be well constrained from the frequency dependence of the MT *transfer functions*. Surface-wave studies can provide a similar depth control because of the period dependence of the group velocity (e.g., Debayle & Kennett, 2000), but most studies of continental seismic anisotropy use shear-wave splitting results (e.g., splitting of SKS waves), which provide inherently poor control on the depth in which anisotropy occurs.

The classical seismological tool for detecting mantle heterogeneities is tomography. P-wave tomography (Ritter *et al.*, 2001), S-wave tomography (Keyser *et al.*, 2002) and array electromagnetics (Leibecker *et al.*, 2002) have been applied in the same target area in the western Rhenish Massif. Ritter *et al.* (2001) found a P-wave velocity reduction of 2% or less within a 100-km-wide columnar structure within the upper mantle, which they interpreted as a temperature increase of 150–200 K associated with a mantle plume. Leibecker *et al.* (2002) found an electrically anisotropic structure under the Rhenish Massif with an E–W-trending direction of high conductivity. If partial melt were the dominant electrical *conduction* mechanism, then enhanced conductivites might be expected to be restricted within the columnar structure inferred by Rittter *et al.* (2001), which is not the case. A larger array experiment indicated that electrical anisotropy could extend under most of central Germany, and that the conductance maximum of 50 000 S does not occur under the Rhenish Massif but rather 200 km to the east (Gatzemeier, personal communication). Furthermore, the seismological results place an upper limit of 1% on the melt fraction that might be present within the low-velocity column (Keyser *et al.*, 2002), which is insufficient to explain the high conductance found by Leibecker *et al.* (2002). Other conduction mechanisms such as hydrogen diffusivity or electronic conduction must therefore be invoked.

9.3 EM and geodynamics

Combined with a knowledge of conduction mechanisms (Chapter 8), MT can provide insights into the evolution of the lithosphere. For example, Brown (1994) showed that accretion processes from island arcs to mature continental crust (e.g., Nelson, 1991) can leave high-conductivity zones as fossil remnants of collisions. We have seen that electrical conductivities within the Earth can be affected by small quantities of fluids (brines and partial melts) and graphite. The amount of fluid present at different depths within the Earth could have important rheological consequences, given the effects of hydro-lytic weakening. Graphitisation is promoted by shear stress (Ross and Bustin, 1990), and its presence may therefore be indicative of shear deformation. Although it is generally not possible to distinguish unequivocally between conduction mechanisms, or to rule out possible contributions from a combination of conduction mechanisms based solely on the conductance of an anomaly, additional consider-ations (e.g., petrological data) allow us to make educated guesses about the nature of conductors at depth. For example, the presence of extensive networks of interconnected aqueous fluids in stable lower continental crust is generally opposed by petrological data (Yardley, 1986), the geotherm does not cross the solidus, and graphite might therefore be the preferred candidate to explain high conductances in stable deep crust (e.g., Mareschal et al., 1992). On the other hand, in 'young' or active tectonic regions a replenishable supply of fluids might reasonably exist (e.g., Li et al., 2003). Hence, electro-magnetic studies can provide constraints on deformation mechan-isms, locate plate boundaries and shear zones, and detect mantle plumes. Ocean-bottom data from the Society Islands plume are discussed in Section 9.4.

The upper mantle is composed of 60–70% olivine. Geodynamic models of mantle flow have been used to infer that progressive simple shear of the mantle imparted by the drag of an overriding tectonic plate favours alignment of olivine [100] axes parallel to the direction of plate motion (McKenzie, 1979; Ribe, 1989). Hydrogen diffusivities linearly enhance the ionic conductivity of olivine crys-tals as described by the Nernst–Einstein equation (Karato 1990):

$$\sigma = fDcq^2/k_B T, \qquad (9.1)$$

where σ is the electrical conductivity, f is a numerical (correlation) factor approximately equal to unity, D is the diffusivity, c is the concentration of diffusing ions, q is the electrical charge of the charged species, k_B is the Boltzmann constant and T is absolute temperature. Hydrogen diffuses along different axes of olivine crystals at rates that are anisotropic (Mackwell and Kohlstedt, 1990; Kohlstedt and Mackwell, 1998), and this leads to single-crystal electrical anisotropies of the order of 40 between [100] and [001] axes and 20 between [100] and [010] axes. Therefore, if olivine [100] axes have a tendency to be aligned in the direction of plate motion, then the upper mantle might be expected to be electrically anisotropic with the direction of highest conductance (*electromagnetic strike* direction) aligned with the plate-motion direction.

Recently, studies aimed at investigating electrical anisotropy at the base of the lithosphere or in the sub-lithospheric upper mantle have been conducted in Australia (Simpson, 2001b), Scandanavia (Bahr and Simpson, 2002) and Europe (Bahr and Duba, 2000, Leibecker *et al.*, 2002). As discussed in Section 9.2, the conductance of 20 000 S found at the base of the lithosphere below the Rhenish Massif (Leibecker *et al.*, 2002) cannot be explained in terms of a partial-melt model, but could be explained with a model of aligned 'wet' olivine. However, neither partial melt nor aligned 'wet' olivine appear to explain the degree of anisotropy detected in some MT field studies (Simpson, 2002a), implying that the current model is incomplete.

Perhaps the most challenging subject for EM could be a quantitative contribution to the question of how much water the Earth's mantle contains. Subduction is a transport process for oceanic sediments (e.g., Morris *et al.*, 1990) and, therefore, for water: suppose all subduction fronts have a combined length of 20 000 km and the average plate velocity is 3 cm/year. If oceanic sediments are 1 km thick and have a *porosity* of 50%, then the sediments subducted every year contain 3×10^8 m^3 seawater. Most of this water is, of course, already released in the accretionary prism, and another fraction of the water plays a role in back-arc volcanism – possibly both water and melt are monitored when MT is applied to such target areas (e.g., Brasse, *et al.*, 2002; Schilling *et al.*, 1997). But how much water is **not** released in these processes, and becomes part of the convecting mantle? Hydrogen diffusivity is a possible mechanism that can contribute to both the enhanced mantle conductivity and the anisotropy of the mantle conductivity (e.g., Lizzaralde *et al.*, 1995). If it would be possible to separate the effects of hydrogen diffusivity, melt and electronic conduction mechanisms on the bulk conductivity of the mantle, then the conductivity increase due to hydrogen diffusivity could readily be converted into a volume fraction of hydrogen using the Nernst–Einstein equation (Equation (9.1)) and the hydrogen diffusivities of mantle minerals (Wang *et al.*, 1996; Kohlstedt and Mackwell, 1998; Woods *et al.*, 2000; Stalder and Skogby, 2003).

With respect to mantle convection, there is a conflict between the geochemical and geodynamical viewpoints: in the geodynamic model of mantle convection the material in the mantle becomes mixed – why, then, does it not mix those geochemically different reservoirs that feed mid-ocean ridge basalts and plumes (e.g., Matsumoto *et al.*, 1997)? Given that these two reservoirs are most likely contained within the upper and lower mantle, we might address this geodynamic–geochemical dichotomy by considering the possible role of the transition zone in hindering or supporting vertical mass transport. A simple question might then be, 'Are geophysical images of the transition zone the same everywhere?'. A comparison of conductivity models and laboratory measurements of different possible mantle materials (Xu *et al.*, 2000) allows the composition of the mantle in the transition zone to be constrained. Such constraints suggest that compositional anomalies in this depth range are either absent or small compared to the large volumes that EM can resolve. However, MT data with small confidence intervals at sufficiently long periods to image the transition zone are rare.

9.4 The oceanic mantle: ocean-bottom and island studies

Filloux (1973, 1987) and Constable *et al.* (1998) have described instrumentation for ocean-bottom electromagnetic studies. A schematic representation of an ocean-bottom MT instrument is shown in Figure 9.1. The instrument sinks to the ocean floor under the weight of a 60 kg concrete anchor that is fastened below its centre. Electric dipole cables are encased in 5 m lengths of 5 cm diameter polypropylene pipes, which terminate with Ag–AgCl electrodes. Different coded sequences of timed pings allow commands to be transmitted to the instrument whilst on the ocean floor. One such command code allows the user to instruct the instrument to release the anchor. On receiving the relevant command code, a stainless-steel burn wire that connects the anchor to the system is electrolysed away by means of a voltage supplied by internal batteries, and the instrument rises to the surface aided by glass flotation spheres.

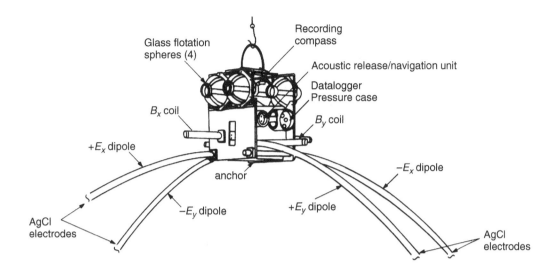

Figure 9.1 Schematic drawing of ocean-bottom MT instrument. The electric dipole arms are 5 m lengths of 5 cm diameter polypropylene pipes. Dipole cables run along the insides of the tubing. (Redrawn from Constable *et al.*, 1998.)

The magnetic variational field at the ocean bottom is attenuated by the conductance of the ocean as described in Section 10.4 (cf. discussion of Equation (3.2) in Section 3.1.1, also). Owing to the skin effect, the attenuation is stronger for shorter periods: the ocean behaves as a low-pass filter. Hence passive ocean-bottom experiments have generally been band-limited to periods longer than 1 minute, and crustal structures have rarely been resolved. A recent attempt to resolve partial melt in the crust and shallow mantle below the East Pacific Rise is reported by Key and Constable (2002).

It has been suggested that differences in the resistivity of oceanic and continental mantle might be expected owing to differences in chemical composition (e.g., Jordan, 1978). Conductivity–depth models based on recent ocean-bottom measurements in stable oceanic regions of the NE Pacific Ocean (Lizzaralde *et al.*, 1995) and Tasman Sea (Heinson and Lilley, 1993) both indicate resistivities in the 5–100 Ω m range in the depth interval 100–400 km, whereas models based on long-period electromagnetic data from the Canadian Shield indicate resistivities in the 15–2000 Ω m range (e.g., Schultz *et al.*, 1993). However, comparisons between the electrical conductivities of oceanic and continental mantle are inconclusive owing to the small number of regions for which data are available, and to the expected heterogeneity of the mantle. Comparison of MT data acquired on Mediterranean islands, which are located in the vicinity of oceanic or transitional mantle, with data from stable continental Europe revealed no significant differences in upper-mantle conductivities in the 100–400 km depth range (Simpson, 2002b).

Acquisition of ocean-bottom data is logistically difficult and expensive. On the other hand, as far as the inductive attenuation of electromagnetic fields is concerned, at periods long enough to penetrate through the seawater layer, measurements on small islands should be approximately equivalent to measurements made on the ocean bottom, since electromagnetic induction is governed by a *diffusion equation*, and what results is a volume sounding, not a point sounding. (Notice the similarity between the complex attenuation factor (Equation (10.13)) at the ocean bottom and the complex reduction of the *Schmucker–Weidelt transfer function* (Equation (2.40)) assuming that an hypothetical instrument swims on the surface of an ocean with conductance τ.) However, island-based electromagnetic studies are made more complicated by distortion of electromagnetic fields at the interface between the relatively resistive island and highly conductive seawater.

Bathymetric effects can be inductive or galvanic, and distortion of magnetic as well as electric fields can occur. For long-period electromagnetic measurements, the *electromagnetic skin depth* is large compared with the seawater depth, and the seawater layer can, thus, be approximated as a thin sheet of equivalent conductance (Section 6.2). The magnetic distortion can then be formulated by considering the magnetic perturbation tensor, $\underline{\underline{W}}$ (Schmucker, 1970; Siemon, 1997), which we describe in more detail in Section 10.1, and which links the magnetic field at a local site to the magnetic field at a reference site. If we identify the reference site with a normal site located outside of the oceanic region that is specified in the thin-sheet model, then Equation (10.1) also describes the distortion at the island site located within the synthesised ocean. The corrections to the horizontal magnetic field are thus given by:

$$\mathbf{B}_{\mathrm{c}} = \left[\underline{\underline{W}}_2 + \underline{I}\right]^{-1}\mathbf{B}_{\mathrm{o}}\ , \tag{9.2}$$

where $\mathbf{B}_{\mathrm{o}}$ and $\mathbf{B}_{\mathrm{c}}$ are the observed and corrected horizontal magnetic fields, $\underline{\underline{W}}_2$ is the 2×2 perturbation tensor consisting of the horizontal components in Equation (10.1) and $\underline{I}$ is the identity matrix.

A similar treatment of the electric fields allows the undistorted *impedance tensor* $\underline{\underline{Z}}_{\mathrm{c}}$ to be calculated from the observed impedance tensor $\underline{\underline{Z}}_{\mathrm{o}}$:

$$\underline{\underline{Z}}_{\mathrm{c}} = \left[\underline{\underline{W}}_2 + \underline{I}\right]\underline{\underline{Z}}_{\mathrm{o}} - \underline{\underline{W}}_2\,\underline{\underline{Z}}_{\mathrm{T}} \tag{9.3}$$

with $\underline{\underline{Z}}_{\mathrm{T}}$ being the impedance due to the thin sheet at the location of the island site and $\underline{\underline{W}}_2$ the modelled horizontal perturbation tensor. Since the correction derived is dependent not only on the thin sheet, but also on the underlying resistivity structure, iteration may be necessary if a significant difference is revealed between the resistivities assumed for the mantle during initial correction of the impedance tensor and the conductivity–depth profile modelled subsequently from the corrected impedance tensor. This approach to correct island MT data for bathymetric effects was applied to data acquired on the Mediterranean islands of Mallorca and Montecristo (Simpson, 2002b). A similar approach was used by Nolasco *et al.* (1998) for removing the bathymetry effects from ocean-bottom data from the Tahiti hotspot. The corrected ocean-bottom data across the Tahiti hotspot are compatible with a modelled mantle plume of limited extent (less than 150 km radius) and slightly higher conductivity than the surrounding mantle (Figure 9.2). Beneath the most active zone, the asthenosphere is laterally heterogeneous, and is more resistive than elsewhere, possibly owing to depletion of volatiles. A possible depression of the olivine–spinel transition (manifest as a significant decrease in resistivity) from 400 to 450 km is also imaged. The data provide no evidence for strong thermal influence over an extensive area, and therefore do not support a superswell model (unless the electromagnetic signature of the superswell is only present in the lower or mid mantle, in which case electromagnetic data with periods longer than 1 day would be necessary to find it). Nor is there evidence for lithospheric thinning owing to plume–lithosphere interaction.

If magnetometers operate simultaneously on the ocean floor and on an island, then a vertical gradient technique (Section 10.4)

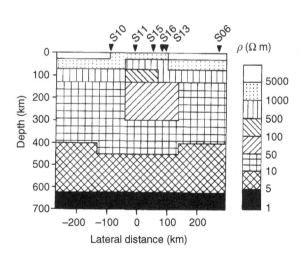

Figure 9.2 Best-fit 2-D model obtained from ocean-bottom MT data over the Tahiti hotspot, French Polynesia, following correction for bathymetric and island effects using 3-D thin-sheet modelling. The electromagnetic strike direction is N130°E, which is in approximate agreement with the spreading direction (N115 ± 5°E) of the Pacific plate. Station S06 is a reference site located ~350 km northwest of the leading edge of the swell. (Redrawn from Nolasco *et al.*, 1998.)

can be applied to infer the vertical conductivity distribution. This technique is particularly useful if telluric data are not available.

9.5 Oceanographic applications

Electromagnetic ocean-bottom measurements have not only been performed in order to measure the conductivity of the oceanic mantle, but also for monitoring oceanic motional fields. As a direct consequence of Faraday's Law (Equation (2.6(a))), when seawater movements with a velocity field, $\mathbf{v}$, interact with the magnetic main field $\mathbf{B}$, of the Earth, they produce a motional electric field:

$$\mathbf{E} = \mathbf{v} \times \mathbf{B} = \begin{pmatrix} \mathbf{x} & \mathbf{y} & \mathbf{z} \\ v_x & v_y & v_z \\ B_x & B_y & B_z \end{pmatrix}, \tag{9.4}$$

where $\mathbf{x}$, $\mathbf{y}$ and $\mathbf{z}$ are unit vectors.

In geomagnetic co-ordinates, $B_y = 0$, and in most cases the vertical component, v_z, of the velocity field is small compared to the horizontal velocity. Therefore Equation (9.4) reduces to (Filloux, 1987):

$$\mathbf{E} = \mathbf{v} \times \mathbf{B} = \begin{pmatrix} \mathbf{x} & \mathbf{y} & \mathbf{z} \\ v_x & v_y & 0 \\ B_x & 0 & B_z \end{pmatrix} = \begin{pmatrix} v_y B_z \\ v_x B_z \\ v_y B_x \end{pmatrix} \tag{9.5}$$

This variational electric field is superimposed on the one arising from induction by external (ionospheric and magnetospheric) sources. This superimposition offers a challenging opportunity: if the impedance describing the induction due to external sources and the variational magnetic field is known, then the electric field due to induction by external sources (in Equation (2.50)) can be removed from the observed electric field, allowing the horizontal water movement to be monitored using Equation (9.5). On the other hand, the electric field created by induction due to internal sources (Equation (9.5)) is regarded as noise if ocean-bottom MT is performed, and in this case the motional electric field has to be removed using statistical or robust procedures (e.g., Section 4.2). The strongest contributions from internal sources are water motions driven by gravitational tides. Gravitational tides have well-known periods and can therefore be removed easily from the observed signal in the frequency domain. A tidal signal is clearly visible in the E–W component of the electric field of ocean-bottom data registered off the N–S-trending coastline of Oregon (Figure 9.3; Bahr and Filloux, 1989).

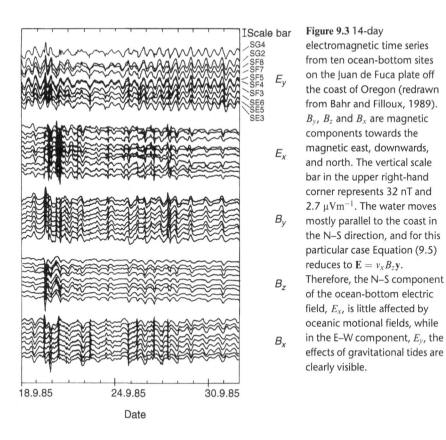

18.9.85   24.9.85   30.9.85

Date

Figure 9.3 14-day electromagnetic time series from ten ocean-bottom sites on the Juan de Fuca plate off the coast of Oregon (redrawn from Bahr and Filloux, 1989). B_y, B_z and B_x are magnetic components towards the magnetic east, downwards, and north. The vertical scale bar in the upper right-hand corner represents 32 nT and 2.7 μVm^{-1}. The water moves mostly parallel to the coast in the N–S direction, and for this particular case Equation (9.5) reduces to $\mathbf{E} = v_x B_z \mathbf{y}$. Therefore, the N–S component of the ocean-bottom electric field, E_x, is little affected by oceanic motional fields, while in the E–W component, E_y, the effects of gravitational tides are clearly visible.

9.6 Industrial applications and environmental studies

MT can be used to map a variety of natural resources including:

(i) oil and gas reservoirs;
(ii) geothermal resources;
(iii) groundwater; and
(iv) ore bodies.

Meju (2002) presents a number of case studies involving the use of EM methods (including, but not limited to MT) for exploring natural resources. Many more commercially sponsored MT surveys have been conducted than have been published. This is partly because industrial partners generally wish to protect their proprietary interests, and partly because using MT for commercial prospecting generally relies on applying existing methodology, rather than driving scientific innovation. The development of marine MT instrumentation can be seen as an exception to the latter statement. The period-dependent attenuation of electromagnetic fields by the ocean (see Sections 9.4 and

10.4) has generally limited academic marine MT research to sounding periods longer than 1000 s, precluding crustal studies. However, the potential to exploit the MT method for assessing oil and gas reserves below salt domes on the ocean bottom has helped to drive the development of a marine MT system that is sensitive to shorter-period signals suitable for imaging shallow structures (Constable *et al.*, 1998). In the low-electrical-noise environment of the ocean bottom, electric fields can be pre-amplified by many orders of magnitude without risk of instrument saturation owing to cultural noise of the type experienced when performing MT on land. The ocean-bottom system described by Constable *et al.* (1998) consists of AC-coupled sensors and pre-amplification factors for the electric fields of 10^6, and *induction coils* for the magnetic field (Figure 9.1).

MT sounding cannot be used to detect oil or gas (which have resistivities exceeding $10^5 \, \Omega \, \mathrm{m}$) directly, but can help to delineate geological structures that can form hydrocarbon traps. The base of a hydrocarbon trap, which is often itself resistive, is frequently marked by a layer of highly conductive sediments or brines, which can be well resolved using MT methods. MT data can be particularly useful for delineating potential hydrocarbon traps below salt domes, volcanics and karsts, which give rise to multiple reflections and scattering of seismic energy. MT can also be adopted for reconnaissance surveys in areas of rugged topography, where large-scale seismic exploration may be prohibitively expensive.

A model study to investigate resolution of subsalt characteristics using marine MT data acquired over salt domes in the Gulf of Mexico has demonstrated that the base of salt structures can be mapped with an average depth accuracy of better than 10% (Hoversten *et al.*, 1998). In this study, the salt dome was assigned resistivities of the order $20 \, \Omega \, \mathrm{m}$, whereas the contacting sediments have resistivities of the order $1 \, \Omega \, \mathrm{m}$. Marine MT surveys can be completed at a fraction of the cost of marine seismic surveys.

Geothermal regions often have enhanced subsurface conductivities owing to hydrothermal circulation (e.g., Fiordelisi *et al.*, 2000). However, geothermal systems incorporate a high level of small-scale 3-D complexity such that lateral resolution of plumbing (i.e., conduits and channels conveying fluids) at depth may be insufficient to obtain more than general constraints on the extent of geothermal prospects (e.g., Bai *et al.*, 2001).

High conductivities modelled below Las Cañadas caldera (Tenerife, Canary Islands), at depths consistent with the depths to the water table measured in boreholes drilled into the caldera, have been interpreted as discontinuous groundwater reservoirs (Pous *et al.*, 2002) that are separated by hydrological barriers. Quantification of the amount of groundwater present is hampered by temperature heterogeneities and unconstrained hydrothermal alteration processes. Integrated EM methods, including AMT, have been used to map groundwater aquifers in the Parnaiba Basin, Brazil (Meju *et al.*, 1999). A reconnaissance AMT survey of shallow sedimentary basins in northern Sudan excluded the presence of exploitable groundwater resources (e.g., Brasse and Rath, 1997).

MT methods can be used to detect massive sulphide deposits and copper deposits if, as is often the case, the host rocks are more resistive than the ore body and the ore is distributed in interconnected veins. Although highly electrically conductive ($5 \times 10^7\,\mathrm{S\,m^{-1}}$), gold is more difficult to detect, because of its low-grade presence in typical deposits, and gold prospecting therefore requires an in-depth understanding of the nature of host rocks and mineralisation. The geological environments in which various ore bodies are found are described in Meju (2002).

Environmental applications of MT include:

 (i) space weather;
 (ii) landfill and groundwater contamination problems; and
 (iii) earthquake studies.

Space weather is a term coined to describe spatial and temporal electromagnetic disturbances (e.g., *magnetic storms*) in the ionosphere and *magnetosphere* associated with solar wind–magnetosphere interactions. Magnetic storms can be hazardous to orbiting satellites and spacecraft, cause radio and TV interference, and generate current surges in power lines and gas pipelines. To negate corrosion of steel gas pipelines, a cathodic protection potential is often applied to the pipeline. Geomagnetic variations modify pipe-to-soil contact potentials and, during magnetic storms, cathodic protection of pipelines may be compromised. MT methods can be used to determine adequate levels of protection potentials (Brasse and Junge, 1984). In a more general sense, magnetometer arrays can be used to monitor geomagnetic activity for forecasting purposes (e.g., Valdivia *et al.*, 1996). Figure 9.4 shows an example of enhanced magnetic activity at mid-latitudes owing to a powerful solar flare that impacted the Earth's magnetosphere on 28th October 2003.

MT sounding can be used as an aid to assessing the *porosity* and *permeability* of potential landfill sites, and to monitor leaching and leakage of conductive chemicals in pre-existing landfill sites. Applications of this type are closely linked to hydrological interests.

Figure 9.4 Three-component
magnetic time series recorded
at Göttingen following a solar
flare that impacted the Earth's
magnetosphere on 28th
October 2003. (Courtesy of
Wilfried Steinhoff.)

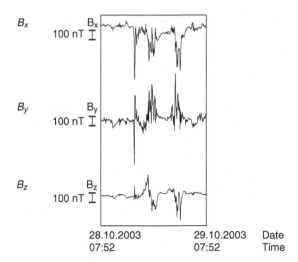

Figure 9.4 Three-component magnetic time series recorded at Göttingen following a solar flare that impacted the Earth's magnetosphere on 28th October 2003. (Courtesy of Wilfried Steinhoff.)

Another source of contamination of freshwater supplies arises from saltwater intrusion. The different resistivities of fresh water and salt water can be used to map the interface between them.

Despite an escalating number of claims that earthquakes might be predicted using EM methods, consideration of the chaotic processes that lead to earthquakes lead us to be sceptical. Magnetotelluric sounding can, however, be used to map faults, and although it is not possible to distinguish between active and passive faults on the basis of conductivity (since both active and passive faults can be conductive for different reasons), knowing the orientation and dip of a major fault with respect to the stress field acting on it may be useful for predicting the potential sense of motion and possible effects of an hypothetical earthquake. Hypothetical damage models incorporating geophysical, geological and civil engineering data are used by insurance companies in order to set insurance premiums in earthquake regions.

Furlong and Langston (1990) suggested a model for an earthquake which nucleated in California at a depth of 18 km in which a discrepancy in directions of displacement in the crust and mantle causes a localised accumulation of strain, which is accommodated by decoupling in the lower crust. This hypothesis might be tested using MT data, since different electromagnetic strikes might be anticipated for the crust and mantle. Thus, the frequency dependence of electromagnetic 'strike', relating to different depth ranges, might potentially offer constraints concerning the disparate motions of the Pacific plate mantle below the north American crust.

Chapter 10
Other EM induction techniques

We finally pay a few other induction techniques a short visit. Magnetometer array studies have been conducted in many parts of the world. Interpretation of the data acquired is mostly restricted to an estimation of the amplitude of the vertical magnetic variational field relative to the amplitude of the horizontal magnetic variational fields, displayed as 'induction vectors'. The interpretation of these data in terms of lateral conductivity contrasts has been rather qualitative, but if regional differences of the horizontal variational fields of different sites are also evaluated, then 'geomagnetic depth sounding' (GDS) provides quantitative information complementary to MT data. Appealingly, the additional information comes at no extra cost: a group conducting an MT campaign with more than one instrument can use the data so acquired for GDS with no additional hardware or field procedures. We demonstrate how the existing processing and interpretation schemes can be modified with only very minor changes. On a more historical note, simultaneous measurements of the magnetic field at the Earth's surface and in boreholes have provided a means of studying the exponential decay of magnetic fields within the Earth (for those who doubt the action of the skin effect).

A fascinating experience is the evaluation of the magnetic daily variation, caused by current vortices in the ionosphere. In this case, the plane wave assumption is violated (the penetration depth – 600 km – is not really very small compared to the size of the ionic current vortices – 6000 km), but this can be turned to advantage. We demonstrate how the MT impedance can be estimated from purely magnetic data. The plane wave assumption is also violated at both polar and equatorial latitudes, imposing both constraints and challenges on the MT method.

We briefly mention the active electromagnetic methods. Books on applied geophysics cover most of these techniques, and we will restrict this section to an appraisal of the advantages and disadvantages of active methods, compared to passive MT methods.

10.1 Magnetometer arrays and magnetovariational studies

Horizontal conductivity contrasts change the direction, amplitude and phase of magnetic variational fields in their vicinity, giving rise to the appearance of *induction vectors* (see Section 2.8). In addition to giving rise to vertical magnetic components, horizontal conductivity gradients also give rise to horizontal changes in the horizontal components of the magnetic field. Quantitatively, regional changes of (anomalies in) the components of the magnetic field can be described with the *perturbation tensor, $\underline{W}$* (Schmucker, 1970, Siemon, 1997), which links the difference $\mathbf{B_a} = \mathbf{B} - \mathbf{B_n}$ between the magnetic field, $\mathbf{B}$, at a recording site and the magnetic field $\mathbf{B_n}$ at a reference (normal) site according to:

$$\begin{pmatrix} B_{a,x} \\ B_{a,y} \\ B_{a,z} \end{pmatrix} = \begin{pmatrix} w_{xx} & w_{xy} \\ w_{yx} & w_{yy} \\ w_{zx} & w_{zy} \end{pmatrix} \begin{pmatrix} B_{n,x} \\ B_{n,y} \end{pmatrix} \qquad \text{or} \qquad \mathbf{B_a} = \underline{\underline{W}}\mathbf{B_n}. \qquad (10.1)$$

The elements, W_{ij}, of the perturbation tensor are dimensionless, complex variables. They can be estimated in the frequency domain from field data with formulae similar to Equations (4.18) and (4.19), by replacing the input process of the bivariate linear system ($\tilde{Z}$ in Equations (4.18) and (4.19)) with $\mathbf{B} - \mathbf{B_n}$, e.g., $\mathbf{B}_x - \mathbf{B}_{n,x}$ for the first component of the vector in Equation (10.1). Egbert (1997) and Soyer and Brasse (2001) have developed multi-site processing schemes. The geomagnetic *transfer functions* can be calculated from modelled magnetic fields with formulae similar to Equation (6.10). An advantage of magnetovariational (MV) studies (sometimes referred to as 'geomagnetic depth sounding' (or GDS)) is that, because no electric fields are measured, the *static shift* problem does not occur. A disadvantage of MV studies is that only horizontal conductivity gradients are determined, which means that the vertical conductivity distribution is not resolved. Therefore, modern applications combine the magnetovariational and the magnetotelluric techniques (e.g., Siemon, 1997). A pre-requisite for the application of the MV technique is the availability of synchronous data from a recording site and a reference site. However, synchronised

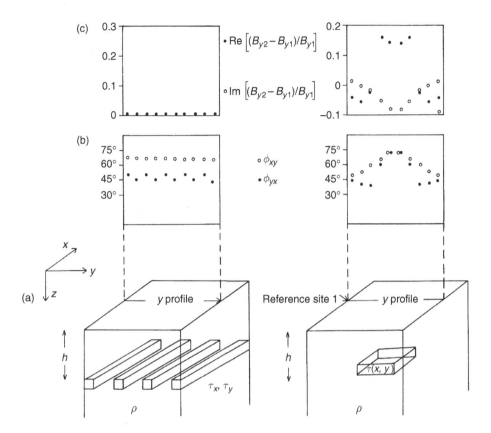

Figure 10.1 Distinguishing between anisotropy and an isolated conductivity anomaly. (a) Two models, one containing macroscopic anisotropy at a depth, *h*, the other containing an isolated conductor at depth, *h*. (b) Impedance phases, ϕ_{xy} and ϕ_{yx}; and (c) GDS transfer functions at a period of 10 s for profiles of sites crossing the anisotropy/isolated conductor.

data recordings are also required for remote-reference estimation of MT transfer functions, which is performed quite routinely.

Whereas *impedance phase* splitting at a single site can be explained either by a model containing anisotropy or one containing an isolated conductivity anomaly, magnetovariational data from an array of sites are different for these two models (Figure 10.1). Hence, magnetovariational studies offer an independent check of the hypotheses involving an anisotropic crust or mantle.

10.2 Equivalence relations revisited in spherical co-ordinates: introduction to induction by the daily magnetic variation and the *Z/H* technique

Unlike the MT technique in which at mid-latitudes a *plane wave* source field can be assumed, the *Z/H technique* (Eckhardt, 1963) utilises the time-variant *solar quiet (Sq)* current vortex with periods

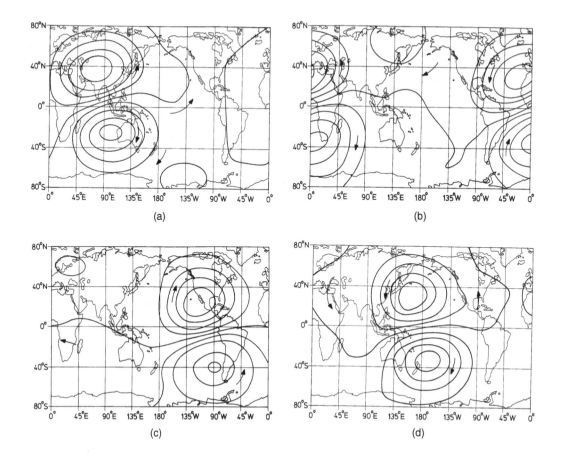

Figure 10.2 Contours of the Sq external current vortex at 6-hour intervals derived from data registered during the September equinox at approximately 50 geomagnetic observatories. (a) 6:00 UT, (b) 12:00 UT, (c) 18:00 UT, (d) 24:00 UT. (Redrawn from Parkinson, 1971.)

of $(24/m)$ hours, where $m = 1, 2, 3, 4$. This current vortex (e.g., Figure 10.2), which does not create a plane wave field, even for a layered (1-D) Earth, gives rise to a vertical magnetic component. This is in contrast to MT soundings made at shorter periods over a layered Earth, for which no vertical component exists. In other words, whereas for periods shorter than $\sim 10^4$ s, a vertical magnetic component indicates horizontal conductivity contrasts within the Earth (Section 2.8), vertical magnetic components (and their associated *induction arrows*) at longer periods are contaminated by source-field geometries.

Complex *penetration depths* (*C-responses*) are derived in the frequency domain as a ratio between the vertical magnetic field (B_z) and one of or both of the horizontal components of the magnetic field (B_h) according to:

$$C = \frac{B_z}{ik B_h},$$
(10.2)

where **k** is the horizontal wavenumber describing the lateral geometry of the source field. In spherical co-ordinates, Equation (10.2) can be expanded as a C-response of degree n and order m:

$$C_n^{(\vartheta)}(\omega_m) = \frac{r_E}{n(n+1)} \frac{dP_n^m}{d\vartheta} \frac{1}{P_n^m} \frac{B_{n,r}^m}{B_{n,\vartheta}^m}$$ (10.3)

$$C_n^{(\lambda)}(\omega_m) = \frac{r_E}{n(n+1)} \frac{im}{\sin\vartheta} \frac{B_{n,r}^m}{B_{n,\lambda}^m},$$ (10.4)

where r_E is the mean Earth's radius, $\omega_m = 2\pi m/(24 \text{ hours})$ is the radial frequency of the mth Sq harmonic, P_n^m is the associated Legendre polynomial, and (r, ϑ, λ) are spherical co-ordinates (with r being the radial component of the geomagnetic field, and ϑ and λ denoting colatitude and longitude, respectively). The P_n^m occur when the Laplace equation is solved in spherical co-ordinates (Campbell, 1987; Schmucker, 1987). Bahr and Filloux (1989) and Simpson (2002b) describe practical applications of the technique with Equations (10.3) and (10.4) solved for the dominant $n = m+1$ terms. The associated Legendre polynomial terms for $n = m+1$ are anti-symmetric about the equator (Figure 10.3), and are most dominant during the equinoxial season. They are given by:

$$P_{m+1}^m(\vartheta) = c_m \sin^m \vartheta \cos\vartheta \Rightarrow \frac{dP_{m+1}^m(\vartheta)}{d\vartheta} = c_m \sin^{m-1}(m\cos^2\vartheta - \sin^2\vartheta). \quad (10.5)$$

Therefore, Equations (10.3) and (10.4) reduce to:

$$C_{m+1}^{(\vartheta)}(\omega_m) = \frac{r_E}{(m+1)(m+2)} \frac{m\cos^2\vartheta - \sin^2\vartheta}{\sin\vartheta\cos\vartheta} \frac{B_r(\omega_m)}{B_\vartheta(\omega_m)}$$ (10.6)

$$C_{m+1}^{(\lambda)}(\omega_m) = \frac{r_E}{(m+1)(m+2)} \frac{-im}{\sin\vartheta} \frac{B_r(\omega_m)}{B_\lambda(\omega_m)}.$$ (10.7)

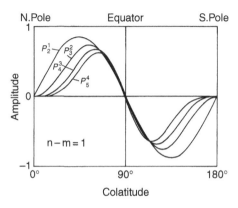

Figure 10.3 Anti-symmetric form about the equator of the amplitude variation as a function of colatitude for associated Legendre polynomial harmonics, $P_n^m, n = m+1$.

The variational magnetic fields can be derived as functions of spherical harmonic coefficients describing internal (arising from induced currents in the Earth) and external (arising from currents in the ionosphere) parts. The relationship between the transfer function C (Equation (10.2)) and the ratio of internal to external parts of the variational field is described in Appendix 2. Spherical harmonic coefficients for internal and external parts of the geomagnetic field have been calculated for four seasons – March equinox, southern summer, September equinox, northern summer – of the International Geophysical Year (1958), (which was a time of sunspot maximum), by Parkinson (1971). The results justify the assumption that the P_{m+1}^{m} term dominates during the equinoxes.

The Z/H technique is best applied to 'quiet days'. These are days that are free from the kind of disturbances associated with *magnetic storms*. Magnetic quiet days can be selected with the help of the Kp index. This is a 3-hourly, quasi-logarithmic scale showing magnetic activity (see, for example, Rangarajan, 1989).

An 8-day segment of simultaneous magnetovariational *time series* for five sites spread over western Europe is shown in Figure 10.4. The Sq daily variation is most clear in the B_y and B_z components. This is because at the latitudes enveloped by the chosen array, the Sq electric currents flow predominantly N–S, such that the eastward component (B_y, or in spherical co-ordinates, the longitudinal component, B_λ) of the Sq magnetic variation dominates significantly over the northward component (B_x, or colatitudinal component, B_ϑ). Figure 10.5 shows spectra for the site 'HEID' in southern Germany. The well-defined peaks are the first four harmonics of the Sq variation.

The problem of describing the time variance of the source field can be avoided by analysing simultaneous segments of data. In this way, the Z/H technique can be applied directly to investigate relative

Figure 10.4 Eight-day segment (14th October 1997–22nd October, 1997) of magnetovariational time series recorded simultaneously at five stations in western Europe: GTTW (Göttingen), HEID (Heidenheim, southern Germany), CLAM (Clamecy, central France), MCHR (Montecristo island), MALL (Mallorca). (After Simpson, 2002b.)

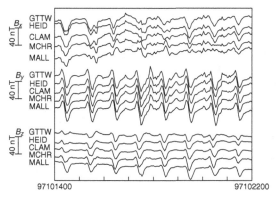

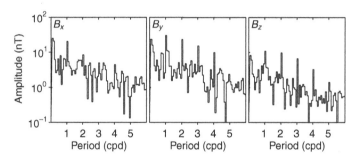

Figure 10.5 Magnetic field spectra at HEID (Heidenheim, southern Germany). The well-defined peaks at 1, 2, 3 and 4 cycles per day (cpd) are the first four harmonics of the *Sq* variation.

differences in mantle conductivities over large regions (Simpson, 2002b). This is much less time-demanding than the more traditional technique of computing average C-responses from coefficients calculated from rigorous spherical harmonic analysis (e.g., Olsen, 1998; Olsen, 1999), and does not average out regional variations in the upper-mantle conductivity structure, which may be of interest. However, C-responses computed directly from short segments of data have fewer *degrees of freedom*, and therefore larger errors than those computed by averaging the coefficients determined from spherical harmonic analysis of a larger dataset.

Whilst most theoretical induction studies with periods of 1 day or longer deal with a layered Earth, Grammatica and Tarits (2002) and Hamano (2002) have attempted to quantify the effects of lateral conductivity variations in a 3-D spherical Earth. Kuvshinov *et al.* (2002) compare the results of theoretical global induction studies involving an heterogeneous Earth with transfer functions derived from geomagnetic observatory data in order to examine the effects of oceans on local C-responses. After correcting the transfer functions from coastal observatories for ocean effects, possible indications of conductivity heterogeneities in the upper mantle are apparent.

An *equivalence relationship* can be shown to exist (Schmucker, 1987) between the MT impedance *Z*, and the C-response:

$$Z = i\omega C. \qquad (10.8)$$

A formal proof of this equivalence relationship is presented in Appendix 2. This equivalence relationship can be used to correct for static shift in long-period MT data (see Section 5.8). The *Z/H* technique can be applied to any passive electromagnetic data that samples several quiet days, provided that the long-period data are not high-pass filtered during recording. The *static shift factor* in MT data has also been determined from magnetovariational data acquired in the vicinity of the equatorial electrojet (see Section

10.3), by applying the Z/H technique to periods shorter than the daily variation, whilst making use of information concerning the geometry of the inhomogeneous source field generated by the equatorial electrojet data (Stoerzel, 1996).

Geomagnetic variational fields with periods longer than 1 day are generated by magnetic storms and are usually described with a P_1^0 geometry, assuming a toroidal ring current outside the Earth. The magnetic field generated by this external ring current (Equation (1.5)) decays within 1–10 days owing to collisions between charge carriers in the *magnetosphere*. Geomagnetic variational fields that are generated by magnetic storms are referred to as Dst (for 'disturbed') data. Combined 1-D modelling of Sq and Dst data (e.g., Banks, 1969; Olsen, 1998) indicates data compatibility with a conductivity–depth profile in which the transition zone (410–660 km) has an average resistivity of 10 Ω m and the lower mantle below 660 km has an average of 1 Ω m (although the usual model non-uniqueness problems apply). The *penetration depths* of Dst data do not exceed 1200 km. Only secular variation studies, which utilise induction due to an internal source (generated by magnetohydrodynamic processes in the Earth's core) provide information about the conductivity below 1200 km (e.g., Aldredge, 1977; Achache *et al.*, 1980).

10.3 The horizontal gradient technique and inhomogeneous sources at high- and low-latitudes

When applying the Z/H technique to the daily variation at the period $T = (24/m)$ hours, we assume that the current vortex in the ionosphere is described by a $\mathrm{e}^{\mathrm{i}m\lambda} P_{m+1}^m (\cos \vartheta)$ spherical harmonic, which dominates the spherical expansion. This approximate description of the source field is most appropriate during the equinoxes when the northern and southern current vortices in the ionosphere are of similar size. If the form of the source field is approximately known, then the horizontal wavenumber, $\mathbf{k}$, can be estimated. Hence Equation (10.2) can be applied to data from a single site. If array data are available, then assumptions about the form of the source field can be avoided by replacing the $\mathbf{k}\mathbf{B}_\mathrm{h}$ term in Equation (10.2) with the horizontal gradient of the horizontal magnetic field:

$$C = \frac{B_z}{\mathrm{i}\left(\dfrac{\partial B_x}{\partial x} + \dfrac{\partial B_y}{\partial y}\right)}. \tag{10.9}$$

Equation (10.9) can be applied to data without the need for a priori knowledge concerning the form of the source field. The horizontal gradient technique (Equation (10.9)) can, therefore, be applied to data affected by inhomogeneous source fields at periods shorter than Sq periods.

Two prominent examples of inhomogeneous source fields that occur at shorter periods than the daily variation are the polar and the equatorial electrojets. The polar electrojet is an highly variable current system in the ionosphere, extending approximately 17° from the poles, and generated by the motion of charged particles from the solar wind that are captured by the Earth's *magnetosphere* and spiral towards the Earth along the magnetic field lines of the cusp region – a gap in the magnetosphere between field lines oriented towards the day-side (towards the Sun) and the night-side of the Earth, respectively. As the charged particles spiral towards the Earth, the magnetic field acting upon them increases, until they are reflected into the opposite hemisphere. The polar electrojet generates magnetic variations in the 1000–10 000 s period range. MT sites stationed at mid-latitudes are sufficiently far away from these current systems that the assumption of a plane wave field is admissible. However, when MT soundings are performed directly under or close to the polar electrojet the *plane wave* assumption is violated.

The equatorial electrojet is a regional amplification of the amplitude of magnetic variations spanning a wide period range owing to a regional increase in the ionospheric conductivity around the geomagnetic equator, where the field lines of the Earth's magnetic field are oriented parallel to the Earth's surface. The equatorial enhancement is not proportional to the strength of the electrojet alone, but also depends on the intensity of the uniform part of the field (Agarwal and Weaver, 1990). Agarwal and Weaver (1990) used thin-sheet modelling to investigate the inductive effect of the daytime electrojet on geomagnetic variations in the region of the Indian peninsula and Sri Lanka. In this study, the electrojet was represented as a Gaussian current distribution with a bandwidth of 600 km and a maximum intensity close to the dip equator, at a height of 110 km. The magnetotelluric impedance was found to be less influenced by inhomogeneous source effects than Z/H values.

Early work on modelling of inhomogeneous sources is reviewed by Mareschal (1986). Most applications of the Z/H technique and the horizontal gradient technique assume a layered Earth. Interpretations of field data using numerical codes (Schmucker, 1995) that allow for both a 2-D Earth and a heterogeneous source

were presented by Bahr and Filloux (1989) and Ogunade (1995).
Jones (1982) applied the horizontal gradient technique (Equation
(10.9)) to the polar electrojet in an early attempt to resolve a sub-
lithospheric conductor under Fennoscandia. However, the existence
of such a conductor was challenged by Osipova *et al.* (1989) who
argued that the drop in the *apparent resistivity* at long periods could
also be a consequence of an inhomogeneous source field. Engels
et al. (2002) used thin-sheet modelling of the conductivity structure
in Fennoscandia to reproduce electromagnetic transfer functions
with an inhomogeneous external current system representing the
polar electrojet.

10.4 Vertical gradient techniques

In those rare cases when a borehole and a borehole magnetometer
are available, the electromagnetic skin effect may be measured by
comparing simultaneous magnetic measurements at the surface of
the Earth and inside the borehole. For a layered Earth, the
x-component of Ampère's Law (Equation (2.6b)) reduces to:

$$-\frac{\partial B_y}{\partial z} = \mu_0 \sigma E_x. \tag{10.10}$$

Therefore, the magnetotelluric impedance can be estimated from
the vertical gradient of B_y according to:

$$Z_{xy} = \frac{E_x}{B_y/\mu_0} = -\frac{1}{\sigma}\frac{\partial B_y}{\partial z}\frac{1}{B_y}. \tag{10.11}$$

Successful application of the vertical gradient technique requires
measurements to be made over a period range in which a detectable
attenuation (i.e., a significant vertical gradient) of the horizontal
magnetic field occurs. Spitzer (1993) provides an example of the
vertical magnetic gradient technique based on borehole magnetometer measurements made to depths of 3 km at the KTB deep
borehole in southern Germany.

Another realisation of the vertical gradient technique involves
simultaneous electromagnetic recordings at sites on the ocean floor
and on the continent (or on one or more islands). If τ is the
conductance of the seawater overlying an ocean-bottom site,
then the *Schmucker–Weidelt transfer functions* for an ocean-
bottom site (C_{ob}) and a land site (C_{cont}) can be related according
to Equation (2.40):

$$C_{cont} = \frac{C_{ob}}{1 + i\omega\mu_0\tau C_{ob}}. \tag{10.12}$$

For long periods, the electric field at the top of the conductor (the ocean) is the same as at the bottom, and therefore the attenuation of the magnetic field by the ocean is:

$$\frac{B_{ob}}{B_{cont}} = \frac{C_{cont}}{C_{ob}} = \frac{1}{1 + i\omega\mu_0\tau C_{ob}}. \tag{10.13}$$

Hence, the transfer function that would be measured with ocean-bottom MT can be obtained from the attenuation ratio:

$$C_{ob} = \frac{\dfrac{B_{cont}}{B_{ob}} - 1}{i\omega\mu_0\tau}, \tag{10.14}$$

without any electric measurements on the ocean bottom. An example is given by Ferguson *et al.* (1990).

10.5 Active induction techniques

Active induction techniques include controlled source AMT (CSAMT), transient electromagnetic (TEM) sounding and long-offset TEM (LOTEM). Detailed descriptions of these methods, including case studies have been presented by Zonge and Hughes (1991) for the case of CSAMT, and Meju (2002) and Strack (1992) for the TEM and LOTEM techniques. We, therefore, restrict this section to a brief appraisal of the advantages and disadvantages of passive MT compared to active induction techniques.

Whereas in the MT technique a uniform source field can be assumed, in active techniques the geometry of the source must be taken into consideration, making mathematical parameterisation and modelling more difficult for the case of active techniques. Therefore, multi-dimensional problems are more easily solved in the case of MT sounding than for active techniques. However, controlled-source techniques can have better resolving power in the limited range that lies between the near-field, for which the source geometry is dominant, and the far-field, for which the source can be approximated as uniform.

TEM can be an efficient mapping technique for shallow structures, because soundings can be made at many sites in a short time. However, the penetration depth that can be achieved with active techniques is limited by the power of the transmitter that provides the active source. In practice, this limits the penetration depth of active techniques to depths within the crust. On the other hand, penetration to depths of the order of the 410 km *transition zone* in the mid mantle are achievable using MT sounding (although the resolution obtained will depend on the regional conductivity structure), with no need of a cumbersome transmitter.

Active induction techniques such as TEM do not have the static shift problems that afflict MT sounding. Indeed, in some environments, TEM data can be used in conjunction with MT data from the same site in order to correct for static shift in MT data (Meju 1996; 2002). However, problems with this approach occur:

(i) Where MT *phase* curves are split at periods that correspond to depths overlapped by TEM data, indicating that differences in resistivities of the two polarisations at high frequency are not only caused by static shift, but also by inductive effects. If the impedance phases are split at frequencies corresponding to depths penetrable by near-surface active measurements, shifting both apparent resistivity curves to a level implied by the active measurements is likely to be incorrect, since any initial difference between the curves is unlikely to be generated purely by static shift, which would not explain the impedance phase splitting.

(ii) Where static shift first affects MT data between the maximum depth of penetration of TEM data and the depth of the target encountered in MT sounding. For example, the use of TEM data to correct for static shift in MT data that is used to image the mantle is often inadequate, because of distortion that arises owing to heterogeneities in the mid to lower crust (e.g., Simpson, 2000).

Appendix 1
Theorems from vector calculus

The gradient, ∇, of a scalar field, ζ, is denoted Grad ζ or $\nabla \zeta$ and describes the rate of change of ζ.

In Cartesian co-ordinates (x, y, z),

$$\nabla \zeta = \left(\frac{\partial \zeta}{\partial x}, \frac{\partial \zeta}{\partial y}, \frac{\partial \zeta}{\partial z} \right) \qquad (A1.1)$$

and in spherical polar co-ordinates (r, ϑ, ϕ)

$$\nabla \zeta = \left(\frac{\partial \zeta}{\partial r}, \frac{1}{r} \frac{\partial \zeta}{\partial \vartheta}, \frac{1}{r \sin \vartheta} \frac{\partial \zeta}{\partial \phi} \right). \qquad (A1.2)$$

The divergence of a vector field, $\mathbf{A}$, is denoted $Div \, \mathbf{A}$ or $\nabla \cdot \mathbf{A}$ and describes the flux of $\mathbf{A}$.

In Cartesian co-ordinates,

$$\nabla \cdot \mathbf{A} = \frac{\partial A_x}{\partial x} + \frac{\partial A_y}{\partial y} + \frac{\partial A_z}{\partial z} \qquad (A1.3)$$

and in spherical polar co-ordinates,

$$\nabla \cdot \mathbf{A} = \frac{1}{r^2} \frac{\partial}{\partial r} \left(r^2 A_r \right) + \frac{1}{r \sin \vartheta} \frac{\partial}{\partial \vartheta} \left(\sin \vartheta A_\vartheta \right) + \frac{1}{r \sin \vartheta} \frac{\partial A_\phi}{\partial \phi}. \qquad (A1.4)$$

The curl of a vector field, $\mathbf{A}$, is denoted Curl $\mathbf{A}$ or $\nabla \times \mathbf{A}$, and describes a rotation of $\mathbf{A}$.

In Cartesian co-ordinates,

$$\nabla \times \mathbf{A} = \left(\frac{\partial A_z}{\partial y} - \frac{\partial A_y}{\partial z}, \frac{\partial A_x}{\partial z} - \frac{\partial A_z}{\partial x}, \frac{\partial A_y}{\partial x} - \frac{\partial A_x}{\partial y} \right) \qquad (A1.5)$$

and in spherical polar co-ordinates,

$$\nabla \times \mathbf{A} = \left(\frac{1}{r \sin \vartheta} \frac{\partial}{\partial \vartheta} \left(\sin \vartheta \, A_\phi \right) - \frac{1}{r \sin \vartheta} \frac{\partial A_\vartheta}{\partial \phi}, \right.$$
$$\left. \frac{1}{r \sin \vartheta} \frac{\partial A_r}{\partial \phi} - \frac{1}{r} \frac{\partial}{\partial r} \left(r A_\phi \right), \frac{1}{r} \frac{\partial}{\partial r} \left(r A_\vartheta \right) - \frac{1}{r} \frac{\partial A_r}{\partial \vartheta} \right) \tag{A1.6}$$

Listed below are some vector identities (e.g., Jackson, 1975). Some of these are used in Chapter 2 as well as in Appendix 2.

$$\mathbf{A} \cdot (\mathbf{B} \times \mathbf{C}) = \mathbf{B} \cdot (\mathbf{C} \times \mathbf{A}) = \mathbf{C} \cdot (\mathbf{A} \times \mathbf{B}) \tag{A1.7}$$

$$\mathbf{A} \times (\mathbf{B} \times \mathbf{C}) = (\mathbf{A} \cdot \mathbf{C})\mathbf{B} - (\mathbf{A} \cdot \mathbf{B})\mathbf{C} \tag{A1.8}$$

$$\nabla \times (\nabla \zeta) = 0 \tag{A1.9}$$

$$\nabla \cdot (\nabla \times \mathbf{A}) = 0 \tag{A1.10}$$

$$\nabla \cdot (\zeta \mathbf{A}) = \zeta \nabla \cdot \mathbf{A} + \mathbf{A} \cdot \nabla \zeta \tag{A1.11}$$

$$\nabla \times (\zeta \mathbf{A}) = \zeta \nabla \times \mathbf{A} - \mathbf{A} \times \nabla \zeta \tag{A1.12}$$

$$\nabla \cdot (\nabla \zeta) = \nabla^2 \zeta \tag{A1.13}$$

$$\nabla \times (\nabla \times \mathbf{A}) = -\nabla^2 \mathbf{A} + \nabla(\nabla \cdot \mathbf{A}) \tag{A1.14}$$

Appendix 2

The transfer function in the wavenumber-frequency domain and equivalence transfer functions

Here, we briefly explain the theory that links MT and the older geomagnetic induction technique that was originally introduced by Schuster and Lamb (Schuster, 1889). At periods of the order of the daily variation and longer, the same *transfer function* can be obtained from MT and from two other induction techniques. Readers interested in the practical aspects of MT might read this section in order to understand how they can obtain the *magnetic reference impedance* from the *magnetic daily variation* (Section 10.2). The *equivalence relationship* that links the transfer functions derived from the different techniques was first recognised by Schmucker (1973), and our presentation follows the unpublished 'Aarhus lecture notes' of Schmucker and Weidelt.

We start with magnetic and electric fields that are derived from a scalar pseudopotential f (which has the dimension [A m]):

$$\mathbf{B} = \mu_0 \, \nabla \times \nabla \times (f\hat{\mathbf{z}}) \tag{A2.1a}$$

$$\mathbf{E} = -\mu_0 \, \nabla \times (\dot{f}\hat{\mathbf{z}}), \tag{A2.1b}$$

where $\hat{z}$ is a unit vector in the vertical direction. $\mathbf{E}$ has only horizontal components, and we restrict the derivation to the case of a layered *half-space*. The fields obey Maxwell's equations (Equations (2.6)) if f follows a *diffusion equation*:

$$\nabla^2 f = \mu_0 \sigma \dot{f}. \tag{A2.2}$$

The measurements are performed in the space-time domain, but transfer functions are derived in the wavenumber-frequency domain.

195

The relationship between these two domains is established by a double *Fourier transform* (see Appendix 5) that links f in the space-time domain to F in the wavenumber-frequency domain:

$$f(x, y, z, t) = \int\limits_{-\infty}^{\infty} \int \int F(z, \mathbf{k}, \omega) e^{i(\mathbf{kr} + \omega t)} dk_x dk_y d\omega, \qquad (A2.3)$$

where $\mathbf{k}$ is a horizontal wavenumber vector such that $|\mathbf{k}| = 2\pi/\lambda$ and $\mathbf{r}$ is a position vector. Notice that the relationship between wavelength and wavenumber magnitude is analogous to the relationship between period and angular frequency. The wavenumber, $\mathbf{k}$, has no vertical component because we only consider the horizontal extension of the source fields, which are generated far away from the surface of the Earth. Transferring the left-hand side of the diffusion equation (Equation (A2.2)) into the wavenumber-frequency domain yields:

$$\nabla^2 f(x, y, z, t) = \int \int \int F(z, \mathbf{k}, \omega) \cdot (-|\mathbf{k}|^2) e^{i(\mathbf{kr} + \omega t)}$$
$$+ \frac{d^2}{dz^2} F(z, \mathbf{k}, \omega) e^{i(\mathbf{kr} + \omega t)} dk_x dk_y d\omega. \qquad (A2.4)$$

As in Section 2.2, Equation (2.2), we introduce an harmonic time factor $e^{i\omega t}$ so that the derivative with respect to time can be expressed as:

$$\mu_0 \sigma \dot{f}(x, y, z, t) = \mu_0 \sigma \int \int \int F(z, \mathbf{k}, \omega) i\omega e^{i(\mathbf{kr} + \omega t)} dk_x dk_y d\omega. \qquad (A2.5)$$

The diffusion equation in the wavenumber-frequency domain is, therefore:

$$\frac{d^2}{dz^2} F(z, \mathbf{k}, \omega) = (|\mathbf{k}|^2 + i\omega\mu_0\sigma) F(z, \mathbf{k}, \omega). \qquad (A2.6)$$

The transfer function in the wavenumber-frequency domain is defined as:

$$\tilde{C}(\mathbf{k}, \omega) = -F(0, \mathbf{k}, \omega) \Big/ \frac{d}{dz} F(0, \mathbf{k}, \omega). \qquad (A2.7)$$

This transfer function has the dimension of length, and describes the decay of fields as they penetrate into the Earth (in the vertical direction). In order to find out how this transfer function is linked to observations made at the surface, we also need to transfer the measured electromagnetic fields into the wavenumber-frequency domain. Using Equation (A1.14) from the theorems of *vector calculus* (presented in Appendix 1), the magnetic field (Equation (A2.1a)) can be expressed as:

$$\mathbf{B} = \mu_0 \nabla \times \nabla \times (f\hat{\mathbf{z}}) = \mu_0 \{\nabla[\nabla \cdot (f\hat{\mathbf{z}}) - \nabla^2(f\hat{\mathbf{z}})]\}$$
$$= \mu_0 \int \int \int [\nabla(\nabla \cdot (F\hat{\mathbf{z}})) - \nabla^2(F\hat{\mathbf{z}})] e^{i(\mathbf{kr} + \omega t)} dk_x dk_y d\omega. \qquad (A2.8)$$

The integral on the right-hand side of Equation (A2.8) can be evaluated:

$$\left[\nabla(\nabla \cdot (F\hat{\mathbf{z}})) - \nabla^2(F\hat{\mathbf{z}})\right] = \nabla\left(\frac{\mathrm{d}F}{\mathrm{d}z}\hat{\mathbf{z}}\right) - \left(-|\mathbf{k}|^2 F\hat{\mathbf{z}} + \frac{\mathrm{d}^2 F}{\mathrm{d}z^2}\hat{\mathbf{z}}\right)$$

$$= \mathrm{i}\mathbf{k}\frac{\mathrm{d}F}{\mathrm{d}z} + \frac{\mathrm{d}^2 F}{\mathrm{d}z^2}\hat{\mathbf{z}} + |\mathbf{k}|^2 F\hat{\mathbf{z}} - \frac{\mathrm{d}^2 F}{\mathrm{d}z^2}\hat{\mathbf{z}}$$

$$= \mathrm{i}\mathbf{k}\frac{\mathrm{d}F}{\mathrm{d}z} + |\mathbf{k}|^2 F\hat{\mathbf{z}},$$

to yield the magnetic field:

$$\mathbf{B} = \mu_0 \iiint \left[\mathrm{i}\mathbf{k}\frac{\mathrm{d}F}{\mathrm{d}z} + |\mathbf{k}|^2 F\hat{\mathbf{z}}\right] e^{\mathrm{i}(\mathbf{kr} + \omega t)} \mathrm{d}k_x \mathrm{d}k_y \mathrm{d}\omega. \qquad (A2.9)$$

Similarly, we can use Equations (A2.1b), (A1.12) and (A2.3) to get the electric field:

$$\mathbf{E} = -\mu_0 \nabla \times (\dot{f}\hat{\mathbf{z}}) = \mathrm{i}\omega\mu_0\hat{\mathbf{z}} \times \nabla f$$

$$= \mu_0 \iiint \mathrm{i}\omega[\hat{\mathbf{z}} \times \mathrm{i}\mathbf{k}F] e^{\mathrm{i}(\mathbf{kr} - \omega t)} \mathrm{d}k_x \mathrm{d}k_y \mathrm{d}\omega. \qquad (A2.10)$$

The fields in the wavenumber-frequency domain are thus:

$$\tilde{\mathbf{B}} = \mu_0 \left[\mathrm{i}\mathbf{k}\frac{\mathrm{d}F}{\mathrm{d}z} + |\mathbf{k}|^2 F\hat{\mathbf{z}}\right] \qquad (A2.11)$$

and

$$\tilde{\mathbf{E}} = \mu_0 \{\mathrm{i}\omega[\hat{\mathbf{z}} \times \mathrm{i}\mathbf{k}F]\}. \qquad (A2.12)$$

The vector product of $\hat{\mathbf{z}}$ and Equation (A2.11) yields:

$$\hat{\mathbf{z}} \times \tilde{\mathbf{B}} = \mu_0\hat{\mathbf{z}} \times \mathrm{i}\mathbf{k}\frac{\mathrm{d}F}{\mathrm{d}z}. \qquad (A2.13)$$

Comparing Equations (A2.12) and (A2.13) we obtain:

$$\tilde{\mathbf{E}} = \mathrm{i}\omega\hat{\mathbf{z}} \times \tilde{\mathbf{B}} \cdot F\left(\frac{\mathrm{d}F}{\mathrm{d}z}\right)^{-1} = -\mathrm{i}\omega\hat{\mathbf{z}} \times \tilde{\mathbf{B}} \cdot \tilde{C}. \qquad (A2.14)$$

Therefore,

$$C_{xy} = \frac{1}{\mathrm{i}\omega}\frac{\tilde{E}_x}{\tilde{B}_y} \quad \text{and} \quad C_{yx} = -\frac{1}{\mathrm{i}\omega}\frac{\tilde{E}_y}{\tilde{B}_x}. \qquad (A2.15)$$

A quick comparison with Equation (2.24) (Section 2.4) confirms that we have reproduced the same basic expression for C in the wavenumber-frequency domain.

Next, we consider two other techniques for obtaining C.

(i) The scalar product of $\hat{z}$ and Equation (A2.11), and the scalar product of $i\mathbf{k}$ and Equation (A2.11) give:

$$\tilde{B}_z = \mu_0 |\mathbf{k}|^2 F \qquad (A2.16a)$$

and

$$i\mathbf{k}\tilde{\mathbf{B}} = -\mu_0 |\mathbf{k}|^2 \frac{dF}{dz}, \qquad (A2.16b)$$

respectively, and from the ratio of Equation (A2.16a) to (A2.16b) we find:

$$C = -F\left(\frac{dF}{dz}\right)^{-1} = \frac{\tilde{B}_z}{i\mathbf{k}\ \tilde{\mathbf{B}}} \qquad (A2.17)$$

Hence, in the wavenumber-frequency domain, the transfer function, C, can be estimated from the ratio of the vertical magnetic component to the horizontal magnetic field gradient $i\mathbf{k}\ \tilde{\mathbf{B}}$, which involves having data available from an array of sites. Alternatively, if the form of the magnetic field, which is characterised by the horizontal wavenumber vector $\mathbf{k}$, is known, then the transfer function can be estimated from magnetic data from a single site (Kuckes, 1973). For the daily variation, $|\mathbf{k}|$ is the inverse radius of the current vortices in the ionosphere (Bahr and Filloux, 1989). For periods shorter than a few hours, there is no vertical magnetic field associated with electromagnetic measurements made over a layered *half-space*, and any vertical magnetic field that is detected indicates an horizontal conductivity contrast (Sections 2.6 and 2.8).

(ii) The link between the transfer function C and Schuster's (1889) separation technique can be understood if we identify the external part, $\tilde{B}_{xe}$, and the internal part, $\tilde{B}_{xi}$, of a horizontal component of $\tilde{B}$ with the amplitudes $-F_0^-$, F_0^+ of a solution of the diffusion equation (Equation (A2.6)):

$$F = F_0^- e^{-kz} + F_0^+ e^{+kz} \qquad (A2.18a)$$

$$\left.\begin{array}{l} \Rightarrow \tilde{B}_{xe} + \tilde{B}_{xi} = \mu_0 i k_x \mathbf{k}(-F_0^- + F_0^+) = \tilde{B}_x \overset{(A2.11)}{=} \mu_0 i k_x \dfrac{dF}{dz} \\[2mm] \tilde{B}_{xe} - \tilde{B}_{xi} = \mu_0 i k_x \mathbf{k}(-F_0^- - F_0^+) = \qquad\quad -\mu_0 i k_x F \end{array}\right\} \quad (A.2.19)$$

From Equation (A2.19), we can derive a transfer function that is related to the ratio of internal to external parts of the magnetic field:

$$C = \frac{1}{k}\frac{\tilde{B}_{xe} - \tilde{B}_{xi}}{\tilde{B}_{xe} + \tilde{B}_{xi}} = \frac{1}{k}\frac{1-S}{1+S}, \qquad (A2.20)$$

where $S(\mathbf{k}, \omega) = \tilde{B}_{xi}(\mathbf{k}, \omega)/\tilde{B}_{xe}(\mathbf{k}, \omega)$.

Appendix 3
Probability distributions

Suppose X is a random variable, x_k $(k = 1, K)$, are realisations (measurements) of X, and $P(x_k < U)$ is the probability that x_k does not exceed an upper limit U. The function:

$$F(u) = P(x < U) \tag{A3.1}$$

is called the probability distribution of X. Obviously, $F(-\infty) = 0$ and $F(\infty) = 1$, and:

$$F(b) - F(a) = P(a \leq x \leq b) \tag{A3.2}$$

is the probability that X is in the interval $[a, b]$. The function:

$$f(u) = \frac{\partial F(u)}{\partial u} \tag{A3.3}$$

is the probability density of X and

$$E\{X\} = \int_{-\infty}^{\infty} xf(x)dx \tag{A3.4}$$

is the expected value. The term:

$$E\left\{(X - E\{X\})^2\right\} =: \sigma_x^2 \tag{A3.5}$$

is the variance of X, and σ_x is the standard variation.

The normal (Gaussian) distribution

A standardised random variable:

$$U = \frac{X - E\{X\}}{\sigma_x} \tag{A3.6}$$

has the expected value $E\{U\} = 0$ and the standard variation $\sigma_u = 1$. The *normal distribution*,

$$F(U) = \frac{1}{\sqrt{2\pi}} \int_{-\infty}^{u} e^{-\hat{u}^2/2} d\hat{u} \qquad (A3.7)$$

has a probability density:

$$f(u) = \frac{1}{\sqrt{2\pi}} e^{-u^2/2}. \qquad (A3.8)$$

From Equation (A3.2), the probability that any realisation lies in the $[-U, U]$ interval is:

$$\beta = F(U) - F(-U) = F(U) - 1. \qquad (A3.9)$$

From the area under the *Gaussian curves* shown in Figure G1, we see that a random realisation will be in the $[-1, 1]$ interval with 68% probability (i.e., we have a standard variation of 1 if we *choose* $\beta = 0.68$), in the $[-2, 2]$ interval with 95% probability and in the $[-3, 3]$ with 99% probability. A random variable that is not standardised but has a distribution with the form of the normal distribution is called a normally distributed random variable.

Samples

The arithmetic mean of a sample of size n is:

$$\bar{x} = \frac{1}{n} \sum_{i=1}^{n} x_n. \qquad (A3.10)$$

The expected value of the mean is the expected value of the random variable X

$$E\{\bar{x}\} = E\{X\}. \qquad (A3.11)$$

The meaning of Equation (A3.11) is that a sample of limited size will produce a mean that is not biased. However, because only the expected values of $\bar{x}$ and X are the same, we cannot expect that $\bar{x}$ is the 'true' mean (the one that we would obtain from a sample of unlimited size). The difference is described by the variance (Equation (A3.5)). The variance of the mean is determined by the error propagation in a sample:

$$\sigma_{\bar{x}}^2 = \frac{1}{n} \sigma_x^2 \qquad (A3.12)$$

but the variance of a sample is defined as:

$$S^2 := \frac{\sum_{i=1}^{n}(x_i - \bar{x})^2}{n-1}. \tag{A3.13}$$

The sample consists of n independent bits of information – the so-called *degrees of freedom*. Estimating the mean (Equation (A3.10)) 'takes' one degree of freedom, therefore we have $(n-1)$ in Equation (A3.13).

The chi-squared (χ^2) distribution function

Suppose $U_1, U_2,..., U_\nu$ are standardised random variables, then a new random variable can be defined:

$$\chi_\nu^2 = \sum_{i=1}^{\nu} U_i^2 \tag{A3.14}$$

describing the chi-squared distribution with ν degrees of freedom:

$$f_{\chi_\nu^2}(U) = \frac{U^{(\nu/2)-1}e^{-U/2}}{\Gamma(\nu/2)\,2^{\nu/2}},$$

where

$$\Gamma(\nu/2) = \int_0^\infty e^{-1}\,t^{(\nu/2)-1}\mathrm{d}t. \tag{A3.15}$$

The expected value of a chi-squared distributed random variable is:

$$E\{\chi_\nu^2\} = \nu, \tag{A3.16}$$

and the variance:

$$\mathrm{var}\left(\chi_\nu^2\right) = 2\nu. \tag{A3.17}$$

The distribution becomes more flat as the number of degrees of freedom increases.

The Fisher F-distribution

Suppose X_1 is a normally distributed random variable with standard variation σ_1. An *estimate* of the variance, estimated from a sample of size n_1, is:

$$S_1^2 = \frac{\sum_{i=1}^{n_1}(x_{1i} - E\{X_1\})^2}{n_1 - 1} \tag{A3.18}$$

Table A3.1 *Upper bounds of the Fisher F-distribution*

ν	5	10	50	100	Infinite
$g_{F_{1,\nu-1}}$ (90%)	4.54	3.36	2.81	2.76	2.71
$g_{F_{1,\nu-1}}$ (95%)	7.71	5.12	4.03	3.94	3.84
$g_{F_{2,\nu-2}}$ (90%)	5.46	3.11	2.41	2.36	2.30
$g_{F_{2,\nu-2}}$ (95%)	9.55	4.46	3.19	3.09	3.00
$g_{F_{4,\nu-4}}$ (90%)	55.8	3.18	2.07	2.0	1.94
$g_{F_{4,\nu-4}}$ (95%)	225.0	4.53	2.58	2.46	2.37

and the new random variable:

$$V_1 = \frac{(n_1 - 1)S_1^2}{\sigma_1^2} \tag{A3.19}$$

is chi-squared distributed with (n_1-1) degrees of freedom. Suppose there is a second normally distributed random variable X_2 with σ_2 and n_2, defining a second chi-squared distributed random variable V_2. The probability distribution of the ratio:

$$F_{n_1-1,\, n_2-1} = \frac{\sigma_2^2 S_1^2}{\sigma_1^2 S_2^2} = \frac{(n_2 - 1)\, V_1}{(n_1 - 1)\, V_2} \tag{A3.20}$$

is the *Fisher F-distribution* with (n_1-1) and (n_2-1) degrees of freedom. With a probability β, it is smaller than its upper bound, $g_{F_{n_1-1,\, n_2-1}}(\beta)$. This upper bound is known for different error probabilities and different degrees of freedom (e.g., Freund and Wilson, 1998). Three cases commonly occur in electromagnetic induction studies:

(i) $F_{1,\nu-1}$ is employed for univariate regression (e.g., Equation (A4.9));
(ii) $F_{2,\nu-2}$ is employed for bivariate regression (e.g., Equation (A4.19));
(iii) $F_{4,\nu-4}$ is employed for bivariate regression with complex input and output processes,

where ν is the sample size. Upper bounds of the Fisher F-distribution for these three conditions are given in Table A3.1.

Appendix 4
Linear regression

A linear system with input X and output Z[11] is described by:

$$Z = AX. \tag{A4.1}$$

Whereas X and Z are random, the parameter A of the linear regression is not. (Electric and magnetic fields can be thought of, with certain restrictions, as random processes – but the *transfer function* (Equation (2.24)) that establishes a linear relationship between them is certainly not random.)

Suppose x_i is a noise-free realisation of X and z_i is a noisy realisation of Z (Figure A4.1). In the following, we use the notation introduced by Gauss:

$$[x] := \sum_{i=1}^{n} x_i. \tag{A4.2}$$

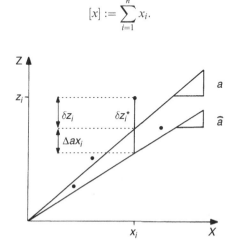

Figure A4.1 Univariate linear regression between an input process, x, and an output process, z. Individual realisation of the input–output process are marked by filled circles. Noise of an individual realisation, δz_i^*, of the output process consists of two terms: individual noise, δz_i, and a term, $\Delta a x_i$, which describes the difference between the estimated regression parameter, a, and the expected value of the regression parameter, $\hat{a}$.

---

[11] In Appendix 4, Z denotes the output process of a linear system and not the MT impedance.

Ideally, we would like to know the expected value (Equation (A3.4)) of the parameter determined from the linear regression: $\hat{a} = E\{A\}$. Instead, we get an estimate a of $\hat{a}$ from a sample of limited size with n realisations. A suggestion is to estimate a by minimising the squared noise term:

$$[(\delta z)^2] = [(z - ax)^2] = min. \tag{A4.3}$$

This minimum is found from the derivative of $[(\delta z)^2]$ with respect to a:

$$\frac{\partial\left[(\delta z)^2\right]}{\partial a} = -2[\delta zx] = 0; \qquad [(z - ax)x] = 0; \qquad [zx] = a[xx] \tag{A4.4}$$

and, therefore,

$$a = \frac{[zx]}{[xx]}. \tag{A4.5}$$

If $z'_i = ax_i$ are the theoretical realisations that are predicted using a, then:

$$\psi^2 = \frac{[z'^2]}{[z^2]} = \frac{[zx]^2}{[xx]\,[zz]} \qquad \text{and } \varepsilon^2 = 1 - \psi^2 \tag{A4.6}$$

are the *coherence* between X and Z and the residuum, respectively. The meaning of coherence can be explained for the special case $\psi^2 = 1$: in this case, $\delta z_i = 0$ or $z'_i = z_i$ for all i – i.e., there is no noise and the output process Z can be predicted entirely from the input process X.

We are also interested in the *confidence intervals* of the parameter of the linear regression, Δa. With respect to the expected value, $\hat{a}$, the noise, δz^*, of an individual realisation of Z consists of two terms: individual noise, δz_i, and a term, Δax, arising from the fact that $a \neq \hat{a}$ (Figure A4.1). Therefore,

$$[\delta z^{*2}] = \{[(z - ax) + (ax - \hat{a}\,x)]^2\} = \left[(\delta z + \Delta ax)^2\right] = [\delta z]^2 + [\Delta a]^2\,[x^2]. \tag{A4.7}$$

The last step made use of the fact that the noise, ΔZ, is not correlated with X. Suppose σ is the variance of δz. We consider three variables with chi-squared probability distributions:

$$\frac{[\delta z^{*2}]}{\sigma^2} = \frac{[\delta z^2]}{\sigma^2} + \frac{(\Delta a)^2\,[x^2]}{\sigma^2} \tag{A4.8}$$

with ν, $(\nu - 1)$ and 1 *degrees of freedom*, respectively. (Estimating Δa takes 1 degree of freedom from the ν degrees of freedom provided by the term on the left-hand side of Equation (A4.8), therefore $(\nu - 1)$ degrees of freedom are remaining for the first term on the right-hand side.). The probability distribution of the ratio:

$$F_{1,\nu-1} = \frac{\nu-1}{1} \frac{(\Delta a)^2 [x^2]}{[\delta z^2]} = \frac{\nu-1}{1} \frac{(\Delta a)^2 [x^2]}{\varepsilon^2 [z^2]} = (\nu-1) \frac{(\Delta a)^2 \psi^2}{\varepsilon^2 a^2} \quad \text{(A4.9)}$$

is a Fisher F-distribution with 1 and $(\nu - 1)$ degrees of freedom. (The last step in Equation (A4.9) made use of Equations (A4.5) and (A4.6)). With the probability:

$$\beta = \int_0^{g_F} f_F(u)du, \quad \text{(A4.10)}$$

the ratio expressed in Equation (A4.9) does not exceed the upper bound $g_{F_{1,\nu-1}}(\beta)$ of the Fisher F-distribution for β with 1 and $(\nu - 1)$ degrees of freedom. Therefore,

$$\frac{(\Delta a)^2}{a^2} \leq \frac{g_{F_{1,\nu-1}}(\beta)}{\nu-1} \frac{\varepsilon^2}{\psi^2}. \quad \text{(A4.11)}$$

From Equation (A4.11), we recognise a few facts which are intuitively clear: firstly, $(\Delta a)^2$ is reminiscent of the variance of a sample (Equation (A3.13)), and we need more than 1 degree of freedom to do statistics *at all*; secondly, if there is no noise, then $a = \hat{a}$ and therefore $\varepsilon^2 = 0$ and $\Delta a = 0$ (from Equation (A3.6)); thirdly, it is not possible to estimate $\hat{a}$ with an error probability $(1-\beta) = 0$, because $g_{F_{1,\nu-1}}(\beta)$ and Δa would be indefinite. We also see from Table A3.1 that for $g_{F_{1,\nu-1}}(\beta)$ the *confidence intervals* provided for $\beta = 95\%$ are larger than those for $\beta = 90\%$.

Bivariate regression

The impedance matrix (Equation (2.50)) establishes two linear systems with two input processes and one output process each. Extending the linear regression to allow for two input processes, we replace Equation (A4.3) by:

$$[\delta z^2] = \left[(z - ax - by)^2\right] = min. \quad \text{(A4.12)}$$

By analogy with Equation (A4.4), the derivatives of the squared residuum with respect to a and b yield:

$$[\delta z \cdot x] = 0 \qquad\qquad [\delta z \cdot y] = 0$$

$$\text{(A4.13)}$$

$$[zx] = a[x^2] + b[yx] \qquad [zy] = a[xy] + b[y^2].$$

The solution of these two equations for two unknowns a, b is:

$$a = \frac{[zx][y^2] - [zy][yx]}{DET}; \qquad b = \frac{[zy][x^2] - [zx][xy]}{DET}, \qquad (A4.14)$$

where

$$DET = [x^2][y^2] - [xy]^2.$$

Considering the coherency between the two input processes:

$$\psi_{xy}^2 = \frac{[xy]^2}{[x^2][y^2]} \qquad (A4.15)$$

we can identify two special cases: (a) $\psi_{xy}^2 = 1$, therefore $DET = 0$, and the bivariate regression fails, because there is only one input process $X + CY$; (b) $\psi_{xy}^2 = 0$, therefore $[xy] = 0$, and the bivariate regression can be replaced by two univariate linear regressions such that Equation (A4.14) reduces to:

$$a = \frac{[zx]}{[xx]}; \qquad b = \frac{[zy]}{[yy]}. \qquad (A4.16)$$

Obviously, the bivariate regression is designed for those cases where the two input processes are partly correlated. Equation (A4.6) is then replaced by the bivariate coherence and residuum:

$$\psi^2 = \frac{[z'^2]}{[z^2]} = \frac{a[zx] + b[zy]}{[z^2]} \qquad \text{and } \varepsilon^2 = 1 - \psi^2. \qquad (A4.17)$$

When estimating the confidence intervals, Δa and Δb, of the parameters of the bivariate regression, we are faced with a minor problem: the analogy of Equation (A4.7):

$$[\delta z^{*2}] = \left[(\delta z - \Delta ax + \Delta by)^2 \right]$$
$$= [\delta z^2] + (\Delta a)^2[x^2] + 2\Delta a \Delta b[xy] + (\Delta b)^2[y^2] \qquad (A4.18)$$

includes a mixed term $\Delta a \Delta b$, and therefore the two confidence intervals cannot be estimated separately. A solution to this problem was provided by Schmucker (1978): in Equation (A4.18), we first replace Δb by $-\Delta a[xy]/[yy]$, in order to estimate Δa. Then, in a second step, we replace Δa by $-\Delta a[xy]/[xx]$, and estimate Δb. Geometrically, the error ellipse is replaced by a rectangle, and the confidence intervals are enlarged. By analogy with Equation (A4.8), we now consider four chi-square distributed variables, but $[\delta z^2]/\sigma^2$ now has $(\nu - 2)$ degrees of freedom, because we have estimated a **and** b. The probability distribution of the ratio:

$$F_{2,\nu-2} = \frac{(\Delta a)^2 \cdot DET \cdot (\nu - 2)}{[y^2] \cdot 2 \cdot \varepsilon^2[z^2]} < g_{F_{2,\nu-2}}(\beta) \qquad (A4.19)$$

is a Fisher F-distribution with 2 and $(\nu - 2)$ degrees of freedom. With a probability β it is smaller than $g_{F_{2,\nu-2}}(\beta)$, and therefore,

$$(\Delta a)^2 \leq \frac{g_{F_{2,\nu-2}}(\beta) \cdot 2 \cdot \varepsilon^2 [z^2] [y^2]}{(\nu - 2) \cdot DET}; \qquad (\Delta b)^2 \leq \frac{g_{F_{2,\nu-2}}(\beta) \cdot 2 \cdot \varepsilon^2 [z^2] [x^2]}{(\nu - 2) \cdot DET}.$$

$$(A4.20)$$

Note that in practical MT, the real parameters a, b of the linear regression are replaced by complex *transfer functions*, and their estimation requires 4 degrees of freedom. Consequently, the confidence intervals of the complex transfer functions depend on $g_{F_{4,\nu-4}}(\beta)$ and the $2/(\nu - 2)$ term in Equation (A4.20) is replaced by $4/(\nu - 4)$. See Table A3.1 for numerical values of the most often occurring values of $g_{F_{4,\nu-4}}(0.95)$.

Confidence intervals of apparent resistivities and impedance phases can be calculated from specified confidence intervals of impedances, ΔZ_{ij}, from:

$$\Delta \rho_{a,ij} = 2(\mu_0/\omega)|Z_{ij}|\Delta Z_{ij} \qquad (A4.21)$$

$$\Delta \phi_{ij} = [\Delta Z_{ij}/\{Re\ Z_{ij}\}][1 + (\{Im\ Z_{ij}\}/\{Re\ Z_{ij}\})^2]^{-1/2}. \qquad (A4.22)$$

Appendix 5
Fourier analysis

The Fourier analysis is described in many textbooks on data processing (e.g., Otnes and Enochson 1972), and we give only a short
outline for those who want to write their own computer algorithm.
In MT, the *Fourier transform* is mostly used to convert *time series*
$E_x(t)$, $B_y(t)$, into the frequency domain, but the same algorithm can
also be applied if we want to convert data from the space domain to
the wavenumber domain (e.g., Appendix 2).

The Fourier transform can only be applied to space or time
series that are stationary – i.e., ones whose probability density
(Appendix 3) is described by a continuous function. (An example
of a non-stationary process is a stock-exchange index – we could not
design a datalogger for monitoring this process over times spanning
a long duration, because no upper limit is known.)

The Fourier transformation of periodic functions is based on
the orthogonality of the trigonometric functions: suppose T is the
period and $\phi(t) = 2\pi t/T$ is a phase angle. The orthogonality theorem states that:

$$
\left.
\begin{aligned}
\int_0^T \cos m\phi \cos m'\phi \, \mathrm{d}t &= \begin{cases} T & m = m' = 0 \\ T/2 & m = m' \neq 0 \\ 0 & m \neq m' \end{cases} \\
\int_0^T \sin m\phi \sin m'\phi \, \mathrm{d}t &= \begin{cases} 0 & m = m' = 0 \\ T/2 & m = m' \neq 0 \\ 0 & m \neq m' \end{cases} \\
\int_0^T \sin m\phi \cos m'\phi \, \mathrm{d}t &= 0
\end{aligned}
\right\} .
\tag{A5.1}
$$

Therefore, if a time series $x(t)$ is a superposition of oscillations of different periods:

$$x(t) = \sum_{m=0}^{\infty} a_m \cos m\phi + b_m \sin m\phi, \quad b_0 = 0, \tag{A5.2}$$

then the coefficients a_m, b_m can be found from:

$$\left. \begin{array}{c} \dfrac{2}{T} \displaystyle\int_0^T x(t) \cos m\phi \; \mathrm{d}t = a_m \\[2mm] \dfrac{2}{T} \displaystyle\int_0^T x(t) \sin m\phi \; \mathrm{d}t = b_m \\[2mm] \dfrac{1}{T} \displaystyle\int_0^T x(t) = a_0 \end{array} \right\}. \tag{A5.3}$$

Parseval's theorem states that information is entirely transferred between the time domain and the frequency domain:

$$\frac{1}{T} \int_0^T x^2(t)\mathrm{d}t = a_0^2 + \frac{1}{2} \sum a_m^2 + b_m^2. \tag{A5.4}$$

Combining the coefficients a_m and b_m to form a complex Fourier coefficient $c_m = a_m - ib_m$, Equation (A5.2) can be replaced by:

$$x(t) = \sum_{m=0}^{\infty} \mathrm{Re}(c_m e^{im\phi}). \tag{A5.5}$$

Naturally occurring processes are generally not periodic. However, if they are stationary the transition $T \to \infty$ is possible. The discrete sequence formed by the Fourier coefficients c_m is then replaced by the Fourier transform, $\tilde{x}(\omega)$ of $x(t)$. With the angular frequency $\omega = 2\pi m/T$ and $m\phi = \omega t$, we then have:

$$x(t) = \frac{1}{2\pi} \int_{-\infty}^{\infty} \tilde{x}(\omega)e^{i\omega t}\mathrm{d}\omega, \quad \tilde{x}(\omega) = \int_{-\infty}^{\infty} x(t)e^{i\omega t}\mathrm{d}t. \tag{A5.6}$$

However, the condition that $x(t) = 0$ must be satisfied at the boundaries.

Finally, we consider digital data. Suppose we have an aperiodic, digital function:

$$x_i = x(i\,\Delta t), \quad i = 1, \ldots, N - 1. \tag{A5.7}$$

We make this function periodic:

$$x_0 = \frac{1}{2}(x(0) + x(T)) = x_n, \quad T = N\Delta t. \tag{A5.8}$$

By analogy with Equation (A5.2), this function is a superposition of oscillations of different periods. Due to the digitisation, the information contents and therefore the number of coefficients is finite:

$$x_j = \sum_{m=0}^{M} (a_m \cos m\phi_j + b_m \sin m\phi_j) + \delta x_i, \qquad (A5.9)$$

where $\phi_j = 2\pi j/N$.

The Fourier coefficients are found with the discrete Fourier transform:

$$\left.\begin{array}{l} a_m = \dfrac{2}{N} \sum\limits_{j=0}^{N-1} x_j \cos m\phi_j \\[3mm] b_m = \dfrac{2}{N} \sum\limits_{j=0}^{N-1} x_j \sin m\phi_j \end{array}\right\} \quad m = 1, ..., M \le N/2$$

$$a_0 = \frac{1}{N} \sum_{j=0}^{N-1} x_j$$

$$a_M = \frac{1}{N} \sum_{j=0}^{N-1} x_j(-1)^j. \qquad (A5.10)$$

Obviously, $T = N\Delta t$ is the longest period, and has the coefficients a_1, b_1. The *Nyquist period* $T_{NY} = 2\Delta t$ is the shortest period. If all oscillations with periods shorter than $2\Delta t$ have been removed from the data prior to digitisation (e.g., with an anti-alias filter, see Section 3.1), then:

$$\tilde{x}(\omega) = 0 \; \forall \omega > \omega_{NY} : = \frac{2\pi}{2\Delta t}. \qquad (A5.11)$$

Because the Fourier transform only exists if $x(t) = 0$ at the margins, we need to prepare the time series prior to the application of Equation (A5.10) in two steps:

(i) we remove a linear trend:

$$x_i \to x_i - i\, a\, \Delta t,$$

where

$$a = \frac{\sum\limits_{i=1}^{N-1} x_i\, i\, \Delta t}{\sum\limits_{i=1}^{N-1} (i\,\Delta t)(i\,\Delta t)} \quad \text{from (A4.5).} \qquad (A5.12)$$

(ii) we multiply x with a *cosine bell*:

$$x_i \rightarrow w_i x_i$$

where

$$w_i = 1/2 + 1/2 \cos \frac{\pi(i + N/2)}{N/2} \quad \text{if} \quad i \leq N/2, \qquad (A5.13)$$

or

$$w_i = 1/2 + 1/2 \cos \frac{\pi(N/2 - i)}{N/2} \quad \text{if} \quad i < N/2.$$

The last step reduces the frequency resolution of the resulting raw spectrum. However, Weidelt's relationship, Equation (2.37), tells us that a high-frequency resolution is not required in MT: neighbouring frequencies yield neighbouring *transfer functions C* and, therefore, similar *apparent resistivities* and *phases*.

Appendix 6
Power and cross spectra

Suppose $x(t)$ and $y(t)$ are stationary processes. The function:

$$c_{xx}(\tau) = \lim_{T\to\infty} \frac{1}{2T} \int_{-T}^{T} x(t)x(t+\tau)\mathrm{d}t \qquad (A6.1)$$

is called the auto-covariance of $x(t)$ at a time-lag τ. The cross-covariance between $x(t)$ and $y(t)$ at a time-lag τ is:

$$c_{xy}(\tau) = \lim_{T\to\infty} \int_{-T}^{T} x(t)y(t+\tau)\mathrm{d}t. \qquad (A6.2)$$

The function:

$$C_{xx}(\omega) = \frac{1}{2\pi} \int_{-\infty}^{\infty} c_{xx}(\tau)\cos(\omega\tau)\mathrm{d}\tau = \frac{1}{\pi} \int_{0}^{\infty} c_{xx}(\tau)\cos(\omega\tau)\mathrm{d}\tau \qquad (A6.3)$$

is the theoretical power spectrum of $x(t)$. The co- and quadrature spectrum or cross-spectrum is:

$$C_{xy}(\omega) = \frac{1}{\pi} \int_{0}^{\infty} c_{xy}(\tau)\cos(\omega\tau)\mathrm{d}\tau. \qquad (A6.4)$$

In summary, theoretical spectra are calculated from a multiplication in the time domain with a lag (Equations (A6.1) and (A6.2), followed by a Fourier transformation into the frequency domain (Equations (A6.3) and (A6.4). Practical estimates of the spectra are calculated in the opposite order: the digital *times series* are first converted into the frequency domain, using Equation (A5.10). If

$$\tilde{X}_m(\omega)\tilde{X}_m^*(\omega) \quad \text{and} \quad \tilde{X}_m(\omega)\tilde{Y}_m^*(\omega) \tag{A6.5}$$

are Fourier products at a particular evaluation frequency, estimates of the power- and cross-spectrum are obtained by averaging these Fourier products:

$$\langle \tilde{X}_m \tilde{X}_m^* \rangle \quad \text{and} \quad \langle \tilde{X}_m \tilde{Y}_m^* \rangle, \tag{A6.6}$$

where $\tilde{X}_m^*$ denotes the complex conjugate of $\tilde{X}_m$.

In contrast to Equations (A6.3) and (A6.4), these averages are limited to a finite data-acquisition time (otherwise, MT would not be practicable). The squared length of this data-acquisition time can be thought of as a linear measure of the number of *degrees of freedom*, which in turn determines the quality of data (see Equation (A4.20)). The practical estimation of a power spectrum and the number of degrees of freedom is outlined in Equation (4.10) and (4.12) (Section 4.1).

Glossary

adjustment length See *horizontal adjustment length*.

aliasing Distortion of a signal caused by undersampling in the time or space domain (e.g., Figure 3.4, Section 3.1.4; Figure 3.7, Section 3.2). See also *Nyquist period*.

apparent resistivity An average resistivity for the volume of Earth sounded by a particular MT sounding period. Apparent resistivity is related to impedance via Equation (2.51). For an homogeneous Earth the apparent resistivity represents the actual resistivity, whereas for a multi-dimensional Earth, the apparent resistivity is the average resistivity represented by an equivalent uniform *half-space*.

aurora australis Luminous phenomenon observed as streamers or bands of intense light in the upper atmosphere in the vicinity of the magnetic south pole caused by excitation (and subsequent decay of excitation state leading to light emissions from excited electrons) of molecules in the Earth's atmosphere owing to collisions with high-energy, charged particles from the solar wind that become entrained along Earth's magnetic field lines, during periods of enhanced magnetic activity. A similar phenomenon (*aurora borealis*) occurs in the northern polar regions.

aurora borealis See *aurora australis*.

backing-off See *compensation*.

Berdichevsky average The arithmetic mean of the principal (off-diagonal) components of the *impedance tensor* (Equation (8.9), Berdichevsky and Dimitriev, 1976).

bivariate linear regression A method of statistical data analysis in which a correlated variable, Y, is estimated from a linear relationship ($Y = aX_1 + bX_2$, cf. Equation (4.15) involving two independent variables, X_1 and X_2, by determining (for example, using the method of *least-square processing*) the straight line that minimises the misfit to the data variables, X_1 and X_2. See also Appendix 4.

B-polarisation For a 2-D Earth, induced electric and magnetic fields are perpendicular to the inducing fields, and the *transfer functions* derived from them can be decoupled into two independent modes. One of these modes is referred to as the **B**-*polarisation*, for which a magnetic field, H_x, parallel to strike is associated with electric fields, E_y, perpendicular to strike and E_z in the vertical plane (as defined in Equation (2.47b) and illustrated in Figure 2.5). This configuration of electric and magnetic fields, for which currents flow perpendicular to strike (y-direction in Figure 2.5) is also referred to in some texts as the *transverse magnetic* or *TM mode*). See also **E**-*polarisation*.

bulk conductivity Average conductivity of a composite medium (for example of a resistive host rock and a conductive electrolyte) often neglecting directional preferences.

calibration Process of estimating and/or applying correction factors to allow for the (frequency-dependent and frequency-independent) performance characteristics of individual sensors (particularly their filters). Section 4.1 describes how calibration is performed in the frequency domain.

coherence, ψ A dimensionless real variable (Equation (3.8)) with values in the range $0 \leq \psi \leq 1$ that is a measure for the amount of linear correlation (1 indicating perfect coherence) between two spectra e.g., electric and magnetic field spectra. See also *confidence interval* and *degrees of freedom*.

compensation The act of applying a stable voltage to the magnetic sensor in order to counterbalance the voltage that is generated by the geomagnetic main field, which has an amplitude that is orders of magnitude greater ($\sim$50 000 nT compared to $<$ 10 nT) than the fluctuations that are of interest in MT studies. Natural self-potentials, which can be larger than induced voltages, are eliminated similarly from measurements of the electrical components.

complex Fourier coefficients Complex (see *complex number*) coefficients derived by Fourier transforming *time series* data into the frequency domain. See Appendix 5 and Equation (4.1).

complex number A number of the form $a + ib$, where a and b are real numbers, and $i = \sqrt{-1}$.

conduction Transmission of electrical charge by the passage of electrons or ionised atoms or molecules.

confidence interval A statistical interval that has a specified probability of containing the predicted value of a parameter. The end points of a confidence interval are referred to as the *confidence limits*. Assuming that errors describe a *Gaussian distribution*, 68% confidence limits lie approximately one standard deviation from the mean, 95% confidence limits lie approximately two standard deviations from the mean and 99% confidence limits lie approximately three standard deviations from the mean (Figure G1).

confidence limits See *confidence interval*.

cosine bell A function (see Equation (A5.13)) that varies in amplitude in proportion to the cosine function during a specified time interval and is zero outside of the specified time interval.

C-response Complex *inductive scale length* computed from variations in the Earth's magnetic field (e.g., Equation (10.2)). See also *Schmucker–Weidelt transfer function* and *equivalence relationship*.

dead-band A frequency range spanning approximately 0.5–5 Hz in which natural electromagnetic fluctuations are of low intensity compared to other frequency ranges in the electromagnetic power spectrum. See, for example, Figure 1.1.

decimation Technique of sub-sampling and low-pass filtering (in order to remove frequencies higher than the *Nyquist frequency*) data so that a *time window* composed of a given number of discrete samples, N, can be processed to yield longer-period *transfer functions* than a *time window* with N undecimated samples. For example, if data are recorded at a sample rate of 2 s, then the *Nyquist period* is 4 s and the longest period present in a *time window* with $N = 512$ is 1024 s, whereas if data are decimated by sub-sampling every fourth point, then the new sample rate is 8 s, the *Nyquist period* is 16 s and the longest period present in a *time window* with $N = 512$ decimated samples is 4096 s (4×1024 s).

declination Angle between magnetic north and true (geographic) north.

decompose To treat the *impedance tensor* in the way described by the *decomposition model*.

decomposition model (hypothesis) A model based on the hypothesis (Larsen, 1975; Bahr, 1988; Groom and Bailey, 1989) that the measured MT *impedance tensor* can be separated into a *distortion matrix* that represents a non-inductive part (often described as *local*), and a modified impedance tensor that represents a *regional*, 1- or 2-D inductive structure. Figure 5.6 summarises the conditions under which the hypothesis is appropriate.

degrees of freedom, v The number of independent observations used to estimate a parameter. The greater the number of degrees of freedom, the smaller will be the *confidence intervals* for the parameter that is estimated (e.g., Equation (4.19)). The number of degrees of freedom for the electromagnetic spectra used to calculate MT response functions is discussed in Section 4.1 (Equation (4.12)). See also Appendix 4.

delta (δ) technique An extension of Bahr's (1988) *superimposition model* to be applied in the case of non-vanishing skew that allows for a *phase* difference (δ) between elements of a given column of an *impedance tensor*, as defined in Equation (5.28).

determinant average A *rotational invariant* formed from the square root of the determinant of the impedance matrix (Equation (8.10)) that represents the geometric average (mean) impedance.

diffusion equation A partial differential equation taking the form shown in Equation (2.13) (i.e., the Laplacian of a quantity, $\nabla^2 E$, is equal to the rate of change of that quantity, $\partial E / \partial t$, at a fixed point in space multiplied by a constant, $\mu_0 \sigma$, that governs waves whose energy is exponentially dissipated.

discrete Fourier transformation (DFT) The formulation of the *Fourier transform* for discrete signals (e.g., Equation (A5.10)).

distortion tensor A matrix containing only real components that represents the non-inductive part of an *impedance tensor* (e.g., Bahr, 1988).

diurnal variation See *magnetic daily variation*.

D^+ model A depth sequence of conductance delta functions (infinitesimally thin sheets of finite conductance) embedded in a perfectly insulating *half-space*, which represents the 1-D model with the smallest misfit with respect to a particular MT *transfer function* (Parker, 1980). See also H^+ model.

Earth response function Characteristic response of the Earth to an input signal. In MT studies, the input signals are natural electromagnetic fluctuations and the Earth's response is typically expressed in the form of an *impedance tensor*.

effective conductivity The *bulk conductivity* of a two-phase medium. For the common case in the Earth's crust that the conductivity of the rock matrix is negligible compared to the conductivity of electrolytes (fluids) or electronic conductors (e.g., ores) contained within it, the *effective conductivity* can be approximated by Equation (8.5) (Waff, 1974). See also *Hashin–Shtrikman upper and lower bounds*.

electromagnetic array profiling (EMAP) An adaptation of the MT technique that involves deploying electric dipoles end-to-end along a continuous profile. The continuous sampling of lateral electric field variations afforded by the technique is sometimes used to suppress *static shift*, as described in Section 5.8.

electromagnetic skin depth The period-dependent (real-valued) depth at which electromagnetic fields are attenuated to e^{-1} of their amplitudes at the surface of the conducting Earth. The attenuation rate depends on the conductivity of the medium penetrated, but longer-period fluctuations are attenuated more slowly than shorter-period fluctuations, and long-period electromagnetic fields therefore penetrate more deeply than short-period electromagnetic fields (Equation (1.1)).

electromagnetic strike Direction of higher conductivity in a 2-D or anisotropic medium; direction of conductive lineaments intersecting a horizontal plane. For a simple 2-D Earth, the strike angle, α, is obtained from the MT *impedance tensor* via Equation (5.13), whereas Equations (5.27) or (5.30) are appropriate if the criteria connected with the *decomposition hypothesis* are fulfilled. See also *phase-sensitive strike*.

E-polarisation For a 2-D Earth, induced electric and magnetic fields are perpendicular to the inducing fields, and the *transfer functions* derived from them can be decoupled into two independent modes. One of these modes is referred to as the E-polarisation, for which an electric field, E_x, parallel to strike is associated with magnetic fields, B_y, perpendicular to strike and B_z in the vertical plane (as defined in Equation (2.47a) and illustrated in Figure 2.5). This configuration of electric and magnetic fields, for which currents flow parallel to strike (*x*-direction in Figure 2.5) is also referred to in some texts as the *transverse electric* or *TE mode*). See also **B**-*polarisation*.

equivalence relationship A relationship (Equation (5.40), Schmucker, 1987), that links the MT *impedance tensor* to the *Schmucker–Weidelt transfer function* or *C-response*, a formal proof of which is given in Appendix 2. This relationship is sometimes used to correct for long-period *galvanic distortion* of MT *transfer functions*, using reference *C-responses* that are derived from purely magnetic fields (see Sections 5.8 and 10.2).

fast Fourier transform (FFT) A type of *discrete Fourier transformation (DFT)* in which the number of discrete samples is normalised to an integer factor of 2 (e.g., $2^9 = 512$), in order to optimise the number of numerical operations necessary in the implementation of the DFT.

fluxgate magnetometer Sensor operating on the principle of hysteresis that is used to measure the direction and intensity of the geomagnetic field. A typical construction is illustrated in Figure 3.1.

Fourier transform (transformation) A function that facilitates the transformation of a signal (*time series*) between the time domain and the frequency domain, where it is represented by complex coefficients belonging to sinusoidal signals that describe the component frequencies and phases of the signal. For a function $f(t)$, the function $\tilde{f}(\omega)$ is given by Equation (A5.6).

galvanic distortion (effects) Non-inductive (frequency-independent) response of the Earth to electric fields, which may distort MT *impedance tensors*. See also *decomposition model (hypothesis)*, *distortion matrix* and *static shift*.

Gaussian distribution A commonly occurring probability density function that is bounded by a bell-shaped curve (Gaussian curve) and has the form:

$$f(x) = \frac{1}{\sigma\sqrt{2\pi}}\exp\left[-\frac{1}{2}\left(\frac{x-\mu}{\sigma}\right)^2\right], \tag{G1}$$

where μ is the mean and σ is the standard deviation (see Figure G1). For $\mu = 0$ and $\sigma = 1$, the Gaussian distribution becomes the *normal distribution*. See also Appendix 3.

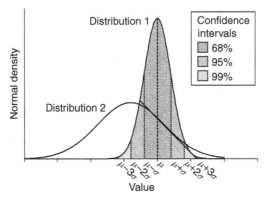

Figure G1 Gaussian curves bounding two Gaussian distributions. Distribution 1 has a mean value of μ and a standard deviation of σ, hence ~68% of values lie in the range $\mu-\sigma$ to $\mu+\sigma$, ~95% of values lie in the range $\mu-2\sigma$ to $\mu+2\sigma$, and ~99% of values lie in the range $\mu-3\sigma$ to $\mu+3\sigma$. These ranges are referred to as confidence intervals. Distribution 2 has a smaller mean value (its peak occurs at a lower value than that of Distribution 1), but a larger standard deviation (wider spread).

Gauss separation Mathematical technique, involving spherical harmonic expansion of the magnetic potential, used for separating internal and external parts of the Earth's magnetic field.

half-space A volume bounded only by an infinite plane (e.g., air–Earth interface, Figure 2.2).

Hashin–Shtrikman upper and lower bounds: Upper and lower bounds on the electrical conductivity of a two-phase medium, as defined by Equation (8.4). Analogous bounds exist for seismic velocities in two-phase media.

horizontal adjustment length The lateral distance (often 2–3 times the *penetration depth*) to which an MT *transfer function* of a given period is sensitive (e.g., Figure 2.8). The exponential decay of current perturbations with lateral distance from the boundary of a conductivity heterogeneity is governed by the conductivity × thickness product of the surface layer and the resistivity × thickness product of the subsurface layer (Equation (2.48); Ranganayaki and Madden, 1980).

hypothesis testing An approach to data interpretation in which a proposed model is investigated for its potential to be acceptable or refutable based on the nearness of its response to observed data. Differs from other approaches to modelling and interpretation of data in that a model that fits the observed data as closely as possible is not necessarily explicitly sought.

H^+ model An extension (Parker and Whaler, 1981) of the D^+ model in which physically unrealistic delta functions are replaced by a stack of homogeneous layers with finite thicknesses and conductivities, terminating at the maximum *penetration depth* of the MT data with a perfect conductor.

impedance phase The phase difference between the electric and magnetic fields used to calculate the MT impedance (see Equations (2.36) and (2.52)). For a uniform *half-space* impedance phases are $45°$, whereas impedance phases higher than $45°$ are indicative of increasing conductivity (decreasing resistivity) with depth, and impedance phases less than $45°$ are indicative of decreasing conductivity (increasing resistivity) with depth. Impedance phases are unaffected by *static shift* and can be used to characterise the inductive part of the *impedance tensor*.

impedance tensor A 2×2 matrix whose components are complex (possessing both magnitudes and phases) ratios of time-varying electric and magnetic fields (Equation (2.50)).

inclination The angle by which the Earth's magnetic field dips from the horizontal plane.

induction arrows Vectorial representations of the complex ratios (Equation (2.49)) of vertical to horizontal magnetic field components that are diagnostic of lateral conductivity variations. See also *Parkinson convention* and *Wiese convention*.

induction coil Type of sensor consisting of an encased coil of copper wire wound onto an elongated high-permeability core that is deployed (usually as a set of three) in MT studies to measure short-period (approximate range 0.001–3000 s) magnetic variations in a particular orientation (e.g., N–S, E–W, vertical). Figure 3.2 summarises the response characteristics (sensitivity) of induction coils versus those of *fluxgate magnetometers*, which are the other types of sensor that are commonly used in MT studies to measure magnetic field variations.

inductive scale length A complex length (e.g., *Schmucker–Weidelt transfer function*, Equation (2.21), for the case of an homogeneous *half-space*) that describes the spatial extent of fields induced in a conductor (the Earth). Compare *penetration depth*.

least count The smallest signal that an analogue-to-digital (A/D) converter can resolve.

least-square processing A statistical processing technique aimed at estimating the best-fit value of a parameter by minimising (in its simplest application by curve fitting) the sum of the squares of the deviations of individual data points from the estimated value (/curve). The success of the technique is dependent on the noise function associated with the data having a *Gaussian distribution*. See also *robust processing*.

least-structure model (inversion) The smoothest (i.e., exhibiting minimal spatial variations as defined by a *roughness function*) model consistent with a specified data misfit. Conceptually attractive because of the inherent diminution in the lateral resolving power of MT data with increasing depth, but nonetheless prone to artefacts, particularly when MT data are anisotropic.

local (response) A non-inductive response owing to multi-dimensional heterogeneities with depths and dimensions significantly less than the *inductive scale length* of an MT *transfer function*. See also *distortion matrix*.

magnetic daily variation Geomagnetic field variation with a period of 1-day, the form of which depends on geomagnetic latitude and local time (see Section 10.2 and Figure 10.2).

magnetic storms Global-scale disturbance of the geomagnetic field that is associated with a sporadic increase in the rate at which plasma is ejected from the Sun, and which is generally characterised by a sudden (time-scale less than one hour) onset and a gradual (time-scale of several days) recovery phase.

magnetopause The outer boundary of the (Earth's) *magnetosphere*, where the solar-wind pressure and the Maxwell pressure (owing to the Earth's magnetic main field) are in dynamic equilibrium.

magnetosphere A region of space surrounding a celestial body (e.g., the Earth) in which ionised particles are affected by the body's magnetic field. The Earth's magnetosphere oscillates in response to increases and decreases in the solar-wind pressure, but extends, on average, to ~10 Earth radii on the dayward side of the Earth and to >40 Earth radii on the Earth's nightward side.

magnetotail Long magnetic tail (>40 Earth radii) of the *magnetosphere* extending from the nightward side of the Earth.

Monte-Carlo technique Automated forward-modelling procedure involving the random sampling of the model space around an input model as a means of investigating the range of model solutions that can fit data to within a specified misfit.

normal distribution See *Gaussian distribution*.

northern lights See *aurora borealis*.

Nyquist frequency Highest frequency (equal to half the sampling frequency) for which information in a digital signal can be unambiguously resolved. See *Nyquist period* and *sampling theorem*.

Nyquist period The threshold period for which signals present in a digital *time series* are adequately resolved. If a *time series* is sampled at intervals of length Δt, then the Nyquist period is $2\Delta t$ (see, for example, Figure 3.4).

Occam inversion Computer algorithm for performing 1-D (Constable *et al.*, 1987) or 2-D inversion (de Groot-Hedlin and Constable, 1990) of MT data that jointly minimises data misfit and model *roughness* over a number of iterations to produce a *least-structure model*.

outlier Value in a dataset that is situated anomalously far from the main distribution of data values.

Parkinson convention Standard practice of drawing *induction arrows* with their arrowheads pointing towards lateral increases in electrical conductivity (Parkinson, 1959).

penetration depth See *electromagnetic skin depth*.

percolation threshold, P_c: The critical value of P for which a connected cluster spans a resistor network from one side to the other, where P is the probability that two sites in a resistor network are connected such that an electric current can flow. The value of P_c depends on the shape and dimension of the network (e.g., Stauffer and Aharony, 1992).

period band A spread of periods bordered by low- and high-pass filters.

permeability Measure of the degree to which a substance (e.g., rock) allows liquids or gases to flow through it. Permeability has the dimensions of an area, m^2.

phase See *impedance phase*.

phase-sensitive skew, η An adhoc measure (Equation (5.32), Bahr, 1988) of the extent to which an *impedance tensor* can be described by the *decomposition model* (Equations (5.22) or (5.28)). **Warning:** For $\eta > 0.3$, the *decomposition model* is inappropriate, but $\eta < 0.3$ does **not** necessarily imply that the *decomposition model* is appropriate!

phase-sensitive strike, α Formulation for the *electromagnetic strike* (Equation (5.27) or Equation (5.30), Bahr, 1988, 1991) based on the premise that an *impedance tensor* satisfies the criteria invoked in the *decomposition hypothesis*.

plane wave A wave with the mathematical form given in Equation (2.2), whose wavefronts (surfaces of constant phase) can be represented by a set of parallel planes.

porosity The ratio (often expressed as a percentage) of the total volume of pore space, which may be filled with fluids, divided by the bulk volume of the host medium (e.g., rock).

pseudosection Contour plot summarising multi-site, multi-frequency data along a profile produced by interpolating a parameter (e.g., **E**-*polarisation* or **B**-*polarisation apparent resistivity* or *impedance phase*) as a function of distance along the profile (usually along the ordinate) and as a function of $\log_{10}$ (frequency) or $\log_{10}$ (period) (usually along the abscissa). (An example of an apparent resistivity pseudosection is shown in Figure 4.7).

Rapid Relaxation (RRI) Inversion Computer algorithm for performing 2-D inversion (Smith and Booker, 1991) of MT data that jointly minimises data misfit and *model roughness* over a number of iterations to produce a *least-structure* model. Uses certain approximations, resulting in a reduction in computation time compared to *Occam inversion*.

raw spectra Unprocessed frequency-domain data (expressed as complex functions).

recursion formula A sequence of formulae or an algorithm that can be used in conjunction with a termination condition that defines f_N (where $_N$ denotes the maximum number of steps in f) to solve iteratively for a value f_n if the value f_{n+1} is known from a previous iteration. An example is Wait's (1954) recursion formula (Equation (2.33)).

regional (response) Inductive part of the MT *impedance tensor*, which is contained in a matrix composed of components that have both magnitude and phase (i.e., its components are complex). Compare *local (response)*.

regularisation mesh A grid within which the differences between some property (e.g., electrical conductivity) of adjacent grid cells is minimised. See also *roughness (function)*.

remote reference processing (method) The use of (usually magnetic) data from a site (denoted as 'remote') located at some distance from the MT site of interest (denoted as 'local') as a means of negating incoherent noise that is uncorrelated between the two sites (Goubau *et al.*, 1979; Gamble *et al.*, 1979, Clarke *et al.*, 1983). Equation (4.17) demonstrates the use of remote reference spectra in solving for the *impedance tensor*.

robust processing A statistical processing technique that utilises iterative weighting of residuals to identify and nullify data biased by non-Gaussian noise (see, for example, Figure 4.3). *Least-square processing* is not robust, because a very small number of strongly biased data can lead to significant bias of the parameter estimated.

rotational (rotationally) invariant Parameter which is unchanged by rotation (i.e., has the same value irrespective of the co-ordinate frame in which it is observed). Examples of rotational invariants are *Swift skew* (Equation (5.16)) and *phase-sensitive skew* (Equation (5.32)).

roughness (function) A penalty function (e.g., Equation (7.6), Smith and Booker, 1991) describing the degree of structural variation present in a model, which is iteratively minimised or reduced to a specified threshold value during *least-structure inversion*.

sampling theorem A time-varying signal can only be adequately recovered from discretely sampled data if the sampling frequency is at least twice the highest frequency component present in the signal (e.g., Figure 3.4). See also *Nyquist period*.

Schmucker–Weidelt transfer function Complex *inductive scale length*, as defined by Equation (2.21) (Weidelt, 1972; Schmucker, 1973). See also *C-response*.

sensitivity The capacity of a sensor/system to respond to input signals, as measured by the ratio of the change in the output that occurs in response to a change in the input. The sensitivity of magnetic sensors is, for example, expressed in $mV\,nT^{-1}$ See Figure 3.2.

skin depth See *electromagnetic skin depth*.

solar quiet (Sq) variations Relatively smooth and predictable form of *magnetic daily variation* (deriving from days when irregular magnetic variations are insignificant or absent) that can be used (together with its harmonics) to calculate vertical to horizontal magnetic field *transfer functions* (*C-responses*: Equations (10.3) and (10.4)) that have *penetration depths* within the Earth of the order 400–600 km.

southern lights See *aurora australis*.

spatial aliasing Error introduced by inadequate sampling (e.g., Figure 3.7) in the space domain (i.e., measurement sites are placed too far apart to adequately resolve small-scale structure). *Static shift* is a common manifestation of spatial aliasing.

spectral line A discrete value (in the frequency-domain) imaged in its most ideal form as a delta function, but commonly represented by a finite spread of values, which corresponds to a prominent frequency in the (electromagnetic) spectrum (e.g., Sq harmonics, Figure 1.1).

spectral matrix A matrix containing power spectra (e.g., $\tilde{X}\tilde{X}^*$, $\tilde{Y}\tilde{Y}^*$) and cross spectra (e.g., $\tilde{Y}\tilde{X}^*$) for a particular evaluation frequency (e.g., Equation (4.11)).

spectral window Interval in the frequency domain over which spectra are averaged and whose form is generally chosen with the aim of optimising the conflicting demands of number of degrees of freedom (which favours including as many data as possible) and data resolution (which favours not smoothing over too many discrete frequencies). See Figure 4.2.

stacking (stacked) A statistical summation of several estimates of a parameter derived from different segments of input data, the aim of which is to increase the *number of degrees of freedom* and therefore the confidence in the parameter estimated. See also *weighted stacking*.

static shift Frequency-independent parallel offset of the magnitudes of *apparent resistivities* that leaves *impedance phases* undistorted (e.g., Figure 5.1). The phenomenon is caused by localised distortion of electric fields by multi-dimensional conductors with depths and dimensions very much less than the *inductive scale lengths* of the effected *transfer functions*. See also *static shift factor*.

static shift factor, s Constant by which *apparent resistivities* are multiplied owing to static shift, which if ignored leads to incorrect determinations of the depth (multiplied by a factor $\sqrt{s}$ in the 1-D case) and conductance (divided by a factor $\sqrt{s}$ in the 1-D case) of conductors. Note: some authors refer to the static shift of the impedance, which in our formalism is $\sqrt{s}$.

superimposition model See *decomposition model (hypothesis)*.

Swift model Simple model (Swift, 1967, reprinted 1986) in which 2-D induction gives rise to an *impedance tensor* that can be rotated into a co-ordinate frame (determined, for example, from Equation (5.13)) such that its diagonal components, Z_{xx} and Z_{yy}, are negligible (in the ideal 2-D case they vanish as in Equation (5.11)) compared to its off-diagonal components, Z_{xy} and Z_{yx}.

Swift skew, κ Ratio (Equation (5.16); Swift, 1967, reprinted 1986) between the magnitudes of the diagonal (Z_{xx} and Z_{yy}) and (principal) off-diagonal (Z_{xy} and Z_{yx}) components of the MT *impedance tensor*, which provides an *ad hoc* measure of the *impedance tensor*'s proximity to an ideal 2-D *impedance tensor* (as approximated by the *Swift model*, Equation (5.11)), for which the diagonal components are expected to vanish. **Warning:** for $\kappa > 0.2$, the *Swift model* is inappropriate, but $\kappa < 0.2$ does **not** necessarily imply that the *Swift model* is appropriate!

Swift strike Electromagnetic strike (Equation (5.13)) obtained from *Swift model*.

time series A set of data sampled at sequential (and usually fixed) time intervals.

time window A segment of data spanning a finite time interval, and composed of a finite number of data points to be analysed.

tipper vectors See *induction arrows*.

transfer function See *C-response, Earth response function, impedance tensor, Schmucker–Weidelt transfer function*.

transient electromagnetic (TEM) sounding An active, time-domain electromagnetic technique for imaging near-surface electrical conductivity structures that involves injecting alternating pulses of electric current into the ground and measuring the voltage registered in a receiving wire loop placed up to a few 100 m from the transmitter.

transition zone(s) Boundary regions situated between the upper and lower mantle characterised by significant increases in density, seismic velocity and electrical conductivity, which are attributable wholly or partly to mineral phase transitions (e.g., olivine to wadsleyite at 410 km).

transverse electric (TE) mode See **E**-*polarisation*.

transverse magnetic (TM) mode See **B**-*polarisation*.

trend removal The filtering of any bias in data that manifests itself as a drift or shift (see, for example, Equation (A5.12), Appendix 5).

vector calculus A branch of calculus in which quantities that can be expressed as vectors are differentiated or integrated. See Appendix 1, for a short introduction.

wave equation A partial differential equation taking the form shown in Equation (2.16) (i.e., the Laplacian of a quantity, $\nabla^2 \mathbf{F}$, is equal to its second partial derivative with respect to time, $\partial^2 F / \partial t^2$, at a fixed point in space divided by the square of the wave velocity, v^2) that governs waves that incur no dissipative attenuation.

weighted stacking An application of stacking that attaches more significance to some data values, and less significance to others, based on a prejudice (see, for example, *robust processing*) concerning the form of the noise distribution function that is imposed on the data.

Wiese convention Non-standard practice (Wiese, 1962) in which *induction arrows* are drawn pointing away from lateral increases in electrical conductivity. See also *Parkinson convention*.

Z/H technique A technique (Eckhardt, 1963; Schmucker, 1987) for estimating complex ratios (Equation (10.2)) between the vertical magnetic field, B_z, and the horizontal components of the magnetic field, B_h, from *solar quiet* (*Sq*) variations. See also *equivalence relationship*.

References

Achache, J., Courtillot, V., Ducruix, J. and LeMouel, J. L. (1980). The late 1960s secular variation impulse: further constraints on deep mantle conductivity. *Phys. Earth Planet. Inter.* **23**: 72–75.

Agarwal, A. K. and Weaver, J. T. (1990). A three-dimensional numerical study of induction in southern India by an electrojet source. *Phys. Earth Planet. Inter.* **60**: 1–17.

Akima, H. (1978). A method of bivariate interpolation and smooth surface fitting for irregularly distributed data points. *ACM Trans. Math. Software* **4**: 148–159.

Aldredge, L. R. (1977). Deep mantle conductivity. *J. Geophys. Res.* **82**: 5427–5431.

Allmendinger, R. W., Sharp, J. W., von Tish, D. *et al.* (1983). Cenozoic and Mesozoic structure of the eastern basin and range province, Utah, from COCORP seismic-reflection data. *Geology*, **11**: 532–536.

Archie, G. E. (1942). The electrical resistivity log as an aid to determining some reservoir characteristics. *Trans. A.I.M.E.* **146**: 389–409.

Bahr, K. (1988). Interpretation of the magnetotelluric impedance tensor: regional induction and local telluric distortion. *J. Geophys.* **62**: 119–127.

Bahr, K. (1991). Geological noise in magnetotelluric data: a classification of distortion types. *Phys. Earth Planet. Inter.* **66**: 24–38.

Bahr, K. (1997). Electrical anisotropy and conductivity distribution functions of fractal random networks and of the crust: the scale effect of connectivity. *Geophys. J. Int.* **130**: 649–660.

Bahr, K., Bantin, M., Jantos, Chr., Schneider, E. and Storz, W. (2000). Electrical anisotropy from electromagnetic array data: implications for the conduction mechanism and for distortion at long periods. *Phys. Earth Planet. Inter.* **119**: 237–257.

Bahr, K. and Duba, A. (2000). Is the asthenosphere electrically anisotropic? *Earth Planet. Sci. Lett.* **178**: 87–95.

Bahr, K. and Filloux, J. H. (1989). Local Sq response functions from EMSLAB data. *J. Geophys. Res.* **94**: 14.195–14.200.

Bahr, K., Olsen, N. and Shankland, T. J. (1993). On the combination of the magnetotelluric and the geomagnetic depth sounding method for resolving an electrical conductivity increase at 400 km depth. *Geophys. Res. Lett.* **20**: 2937–2940.

Bahr, K. and Simpson, F. (2002). Electrical anisotropy below slow- and fast-moving plates: paleoflow in the upper mantle? *Science* **295**: 1270–1272.

Bahr, K., Smirnov, M., Steveling, E. and BEAR working group (2002). A gelation analogy of crustal formation derived from fractal conductive structures. *J. Geophys. Res.* **107**: (B11) 2314, doi: 10.1029/2001JB000506.

Bai, D., Meju, M. and Liao, Z. (2001). Magnetotelluric images of deep crustal structure of the Rehai geothermal field near Tengchong, Southern China. *Geophys. J. Int.* **147**: 677–687.

Banks, R. J. (1969). Geomagnetic variations and the electrical conductivity of the upper mantle. *Geophys. J. R. Astr. Soc* **17**: 457–487.

Bell, D. R. and Rossman, G. R. (1992). Water in the Earth's mantle: the role of nominally anhydrous minerals. *Science* **255**: 1391–1397.

Berdichevsky, M. N. (1999). Marginal notes on magnetotellurics. *Surv. Geophys.*, **20**: 341–375.

Berdichevsky, M. N. and Dimitriev, V. I. (1976). Distortion of magnetic and electric fields by near-surface lateral inhomogeneities. *Acta Geodaet., Geophys. et Montanist. Acad. Sci. Hung.* **11**: 447–483.

Bigalke, J. (2003). Analysis of conductivity of random media using DC, MT, and TEM. *Geophysics* **68**: 506–515.

BIRPS and ECORS (1986). Deep seismic profiling between England, France and Ireland. *J. Geol. Soc. London* **143**: 45–52.

Boas, M. L. (1983). *Mathematical Methods in the Physical Sciences*, 2nd edn. New York: Wiley & Sons.

Brasse, H., Lezaeta, P., Rath, V., Schwalenberg, K., Soyer, W. and Haak, V. (2002). The Bolivian Altiplano conductivity anomaly. *J. Geophys. Res.* **107**: (B5), doi: 10.1029/2001JB000391.

Brasse, H. and Junge, A. (1984). The influence of geomagnetic variations on pipelines and an application of large-scale magnetotelluric depth sounding, *J. Geophys.* **55**: 31–36.

Brasse, H. and Rath, V. (1997). Audiomagnetotelluric investigations of shallow sedimentary basins in Northern Sudan. *Geophys. J. Int.* **128**: 301–314.

Brown, C. (1994). Tectonic interpretation of regional conductivity anomalies. *Surv. Geophys.* **15**: 123–157.

Cagniard, L. (1953). Basic theory of the magnetotelluric method of geophysical prospecting. *Geophysics* **18**: 605–645.

Campbell, W. H. (1987). Introduction to electrical properties of the Earth's mantle. *Pure Appl. Geophys.* **125**: 193–204.

(1997). *Introduction to Geomagnetic Fields*. Cambridge: Cambridge University Press.

Cantwell, T. (1960). Detection and analysis of low-frequency magnetotelluric signals. Ph.D. Thesis, Dept. Geol. Geophys. M.I.T., Cambridge, Mass.

Chamberlain, T. C. (1899). On Lord Kelvin's address on the age of the Earth. *Annual Report of the Smithsonian Institution*, pp. 223–246.

Chapman, D. S. and Furlong, K. P. (1992). Thermal state of the continental crust. In *The Continental Lower Crust*, eds. D. M. Fountain, R. J. Arculus and R. W. Kay. Amsterdam: Elsevier, pp. 81–143.

Chapman, S. (1919). The solar and lunar diurnal variations of terrestrial magnetism. *Phil. Trans. Roy. Soc. London* **A218**: 1–118.

Chapman, S. and Ferraro, V. C. A. (1931). A new theory of magnetic storms. *Terr. Mag.* **36**: 77–97.

Chave, A. D. and Smith, J. T. (1994). On electric and magnetic galvanic distortion tensor decompositions. *J. Geophys. Res.* **99**: 4669–4682.

Clarke, J., Gamble, T. D., Goubau, W. M., Koch, R. H. and Miracky, R. F. (1983). Remote-reference magnetotellurics: equipment and procedures. *Geophys. Prosp.* **31**: 149–170.

Constable, S. C., Parker, R. L. and Constable, C. G. (1987). Occam's inversion: A practical algorithm for generating smooth models from EM sounding data. *Geophysics* **52**: 289–300.

Constable, S. C., Orange, A. S., Hoversten, G. M. and Morrison, H. F. (1998). Marine magnetotellurics for petroleum exploration. Part I: a seafloor equipment system. *Geophysics* **63**: 816–825.

Constable, S. C., Shankland, T. J. and Duba, A. (1992). The electrical conductivity of an isotropic olivine mantle. *J. Geophys. Res.* **97**: 3397–3404.

Davies, G. F. and Richards, M. A. (1992). Mantle convection. *J. Geol.* **100**: 151–206.

Debayle, E. and Kennett, B. L. N. (2000). The Australian continental upper mantle: structure and deformation inferred from surface waves, *J. Geophys. Res.* **105**: 25423–25450.

DEKORP Research Group (1991). Results of the DEKORP 1 (BELCORP-DEKORP) deep seismic reflection studies in the western part of the Rhenish Massif. *Geophys. J. Int.* **106**: 203–227.

Dobbs, E. R. (1985). *Electromagnetic Waves.* London: Routledge & Kegan Paul.

Dosso, H. W. and Oldenburg, D. W. (1991). The similitude equation in magnetotelluric inversion. *Geophys. J. Int.* **106**: 507–509.

Duba, A., Heard, H. C. and Schock, R. N. (1974). Electrical conductivity of olivine at high pressure and under controlled oxygen fugacity. *J. Geophys. Res.* **79**: 1667–1673.

Duba, A., Heikamp, S., Meurer, W., Nover, G. and Will, G. (1994). Evidence from borehole samples for the role of accessory minerals in lower-crustal conductivity. *Nature* **367**: 59–61.

Duba, A. and Nicholls, I. A. (1973). The influence of oxidation state on the electrical conductivity of olivine. *Earth Planet. Sci. Lett.* **18**: 279–284.

Duba, A., Peyronneau, J., Visocekas, F. and Poirier, J.-P. (1997). Electrical conductivity of magnesiowüstite/perovskite produced by laser heating of synthetic olivine in the diamond anvil cell. *J. Geophys. Res.* **102**: 27723–27728.

Duba, A. and Shankland, T. J. (1982) Free carbon and electrical conductivity in the mantle. *Geophys. Res. Lett.* **11**: 1271–1274.

Duba, A. and von der Gönna, J. (1994) Comment on change of electrical conductivity of olivine associated with the olivine–spinel transition. *Phys. Earth Planet. Inter.* **82**: 75–77.

Echternacht, F., Tauber, S., Eisel, M., Brasse, H., Schwarz, G. and Haak, V. (1997). Electromagnetic study of the active continental margin in northern Chile. *Phys. Earth Planet. Inter.* **102**: 69–87.

Eckhardt, D. H. (1963). Geomagnetic induction in a concentrically stratified Earth. *J. Geophys. Res.* **68**: 6273–6278.

Edwards, R. E. and Nabighian, M. N. (1981). Extensions of the magnetometric resistivity (MMR) method. *Geophysics* **46**: 459–460.

Egbert, G. D. (1997). Robust multiple-station magnetotelluric data processing. *Geophys. J. Int.* **130**: 475–496.

Egbert, G. D. and Booker, J. R. (1986). Robust estimation of geomagnetic transfer functions. *Geophys. J. R. Astr. Soc.* **87**: 173–194.

Eisel, M. and Bahr, K. (1993). Electrical anisotropy under British Columbia: interpretation after magnetotelluric tensor decomposition. *J. Geomag. Geoelectr.* **45**: 1115–1126.

Eisel, M. and Haak, V. (1999). Macro-anisotropy of the electrical conductivity of the crust: a magneto-telluric study from the German Continental Deep Drilling site (KTB). *Geophys. J. Int.* **136**: 109–122.

ELEKTB group (1997). KTB and the electrical conductivity of the crust. *J. Geophys. Res.* **102**: 18289–18305.

Engels, M., Korja, T. and BEAR Working Group (2002). Multisheet modeling of the electrical conductivity structure in the Fennoscandian Shield. *Earth, Planets and Space* **54**: 559–573.

Evans, C. J., Chroston, P. N. and Toussaint-Jackson, J. E. (1982). A comparison study of laboratory measured electrical conductivity in rocks with theoretical conductivity based on derived pore aspect ratio spectra. *Geophys. J. R. Astr. Soc.* **71**: 247–260.

Ferguson, I. J., Lilley, F. E. M. and Filloux, J. H. (1990). Geomagnetic induction in the Tasman Sea and electrical conductivity structure beneath the Tasman seafloor. *Geophys. J. Int.* **102**: 299–312.

Filloux, J. H. (1973). Techniques and instrumentation for studies of natural electromagnetic induction at sea. *Phys. Earth Planet. Inter.* **7**: 323–338.

Filloux, J. H. (1987). Instrumentation and experimental methods for oceanic studies. In *Geomagnetism, Volume 1*, ed. J. A. Jacobs. London: Academic Press, pp. 143–248.

Fiordelisi, A., Manzella, A., Buonasorte, G., Larsen, J. C. and Mackie, R. L. (2000). MT methodology in the detection of deep, water-dominated geothermal systems. In *Proceedings World Geothermal Congress,* eds. E. Iglesias, D. Blackwell, T. Hunt, J. Lund and S. Tamanyu. Tokyo: International Geothermal Association, pp. 1121–1126.

Fischer, G. and Le Quang, B. V. (1982). Parameter trade-off in one-dimensional magnetotelluric modelling. *J. Geophys.* **51**: 206–215.

Fischer, G., Schnegg, P.-A., Pegiron, M. and Le Quang, B. V. (1981). An analytic one-dimensional magnetotelluric inversion scheme. *Geophys. J. R. Astr. Soc.* **67**: 257–278.

Freund, R. J. and Wilson, W. J. (1998). *Regression Analysis: Statistical modeling of a response variable.* San Diego: Academic Press.

Frost, B. R. (1979). Mineral equilibria involving mixed volatiles in a C–O–H fluid phase: the stabilities of graphite and siderite. *Amer. J. Sci.,* **279**: 1033–1059.

Frost, B. R., Fyfe, W. S., Tazaki, K. and Chan, T. (1989). Grain boundary graphite in rocks and implications for high electrical conductivity in the crust. *Nature* **340**: 134–136.

Furlong, K. P. and Fountain, D. M. (1986). Continental crustal underplating: thermal considerations and seismic-petrologic consequences. *J. Geophys. Res.* **91**: 8285–8294.

Furlong, K. P. and Langston, C. A. (1990). Geodynamic aspects of the Loma Prieta Earthquake, *Geophys. Res. Lett.* **17**: 1457–1460.

Gaherty, J. B. and Jordan, T. H. (1995). Lehmann discontinuity as the base of an anisotropic layer beneath continents. *Science* **268**: 1468–1471.

Gamble, T. D., Goubau, W. M. and Clarke, J. (1979). Magnetotellurics with a remote magnetic reference. *Geophysics* **44**: 53–68.

Gauss, C. F. (1838). Erläuterungen zu den Terminszeichnungen und den Beobachtungszahlen. In *Resultate aus den Beobachtungen des magnetischen Vereins im Jahre 1837*, eds. C. F. Gauss and W. Weber. Göttingen: Dieterichsche Buchhandlung, pp. 130–137.

  (1839). Allgemeine Theorie des Erdmagnetismus. In *Resultate aus den Beobachtungen des magnetischen Vereins im Jahre 1838*, eds. C. F. Gauss and W. Weber. Göttingen: Dieterichsche Buchhandlung, pp. 1–57.

Glassley, W. E. (1982). Fluid evolution and graphite genesis in the deep continental crust. *Nature* **295**: 229–231.

Goubau, W. M., Gamble, T. D. and Clarke, J. (1979). Magnetotelluric data analysis: removal of bias. *Geophysics* **43**: 1157–1166.

Graham, G. (1724). An account of observations made of the variation of the horizontal needle at London in the latter part of the year 1722 and beginning 1723. *Phil. Trans. Roy. Soc. London* **383**: 96–107.

Grammatica, N. and Tarits, P. (2002). Contribution at satellite altitude of electromagnetically induced anomalies arising from a three-dimensional heterogeneously conducting Earth, using Sq as an inducing source field. *Geophys. J. Int.* **151**: 913–923.

Groom, R. W. and Bahr, K. (1992). Correction for near-surface effects: decomposition of the magnetotelluric impedance tensor and scaling corrections for regional resistivities: a tutorial. *Surv. Geophys.* **13**: 341–379.

Groom, R. W. and Bailey, R. C. (1989). Decomposition of the magnetotelluric impedance tensor in the presence of local three-dimensional galvanic distortion. *J. Geophys. Res.* **94**: 1913–1925.

de Groot-Hedlin, C. (1991). Removal of static shift in two dimensions by regularized inversion. *Geophysics* **56**: 2102–2106.

de Groot-Hedlin, C. and Constable, S. C. (1990). Occam's inversion to generate smooth, two-dimensional models from magnetotelluric data. *Geophysics* **55**: 1613–1624.

Guéguen, Y., David, Chr. and Gavrilenko, P. (1991). Percolation networks and fluid transport in the crust. *Geophys. Res. Lett.* **18**: 931–934.

Guéguen, Y. and Palciauskas, V. (1994). *Introduction to the Physics of Rocks.* Princeton: Princeton University Press.

Haak, V. and Hutton, V. R. S. (1986). Electrical resistivity in continental lower crust. In *The Nature of the Lower Continental Crust,* eds. J. B. Dawson, D. A. Carswell, J. Hall, and K. H. Wedepohl. *Geological Society Special Publication* **24**: 35–49. Oxford: Blackwell Scientific Publications.

Haak, V., Stoll, J. and Winter, H. (1991). Why is the electrical resistivity around KTB hole so low? *Phys. Earth Planet. Inter.* **66**: 12–23.

Hamano, Y. (2002). A new time-domain approach for the electromagnetic induction problem in a three-dimensional heterogeneous Earth. *Geophys. J. Int.* **150**: 753–769.

Hashin, Z. and Shtrikman, S. (1962). A variational approach to the theory of the effective magnetic permeability of multiphase materials. *J. Appl. Phys.* **33**: 3125–3131.

Hautot, S. and Tarits, P. (2002). Effective electrical conductivity of 3-D heterogeneous porous media. *Geophys. Res. Lett.* **29**, No. 14, 10.1029/2002GL014907.

Hautot, S., Tarits, P., Whaler K., Le Gall, B., Tiercelin, J-J and Le Turdu, C. (2000). Deep structure of the Baringo Rift Basin (central Kenya) from three-dimensional magnetotelluric imaging: implications for rift evolution. *J. Geophys Res.* **105**: 23 493–23 518.

Heinson, G. S. (1999). Electromagnetic studies of the lithosphere and asthenosphere. *Surv. Geophys.* **20**: 229–255.

Heinson, G. S. and Lilley, F. E. M. (1993). An application of thin sheet electromagnetic modelling to the Tasman Sea. *Phys. Earth Planet. Inter.* **81**: 231–251.

Hermance, J. F. (1979). The electrical conductivity of materials containing partial melt: a simple model of Archie's law. *Geophys. Res. Lett.* **6**: 613–616.

Hirsch, L. M, Shankland, T. and Duba A. (1993). Electrical conduction and polaron mobility in Fe-bearing olivine. *Geophys. J. Int.* **114**: 36–44.

Hobbs, B. A. (1992). Terminology and symbols for use in studies of electromagnetic induction in the Earth. *Surv. Geophys.*, **13**: 489–515.

Hohmann, G. W. (1975). Three-dimensional induced polarisation and electromagnetic modeling. *Geophysics* **40**: 309–324.

Hoversten, G. M., Morrison, H. F. and Constable, S. C. (1998). Marine magnetotellurics for petroleum exploration, Part II: Numerical analysis of subsalt resolution. *Geophysics* **63**: 826–840.

Huber, P. J. (1981). *Robust Statistics.* New York: John Wiley & Sons.

Huenges, E., Engeser, B., Erzinger, J., Kessels, W., Kück, J. and Pusch, G. (1997). The permeable crust: geohydraulic properties down to 9101 m depth. *J. Geophys. Res.* **102**: 18255–18265.

Hyndman, R. D. and Shearer, P. M. (1989). Water in the lower crust: modelling magnetotelluric and seismic reflection results. *Geophys. J. R. Astr. Soc.* **98**: 343–365.

Jackson, J. D. (1975). *Classical Electrodynamics,* 2nd edn. New York: John Wiley & Sons.

Jenkins, G. M. and Watts, D. G. (1968). *Spectral Analysis and its Applications.* San Francisco: Holden-Day.

Ji, S., Rondenay, S., Mareschal, M. and Senechal, G. (1996). Obliquity between seismic and electrical anisotropies as a potential indicator of movement sense for ductile shear zones in the upper mantle. *Geology* **24**: 1033–1036.

Jödicke, H. (1992). Water and graphite in the Earth's crust – an approach to interpretation of conductivity models. *Surv. Geophys.* **13**: 381–407.

Jones, A. G. (1977). Geomagnetic induction studies in southern Scotland. Ph.D. Thesis, University of Edinburgh.

(1982). On the electrical crust–mantle structure in Fennoscandia: no Moho, and the asthenosphere revealed? *Geophys. J. R. Astr. Soc.* **68**: 371–388.

(1983). The problem of current channelling: a critical review. *Geophysical Surveys* **6**: 79–122.

(1992). Electrical conductivity of the continental lower crust. In *Continental Lower Crust*, eds. D. M. Fountain, R. J. Arculus and R. W. Kay. Amsterdam: Elsevier, pp. 81–143.

(1999). Imaging the continental upper mantle using electromagnetic methods. *Lithos* **48**: 57–80.

Jones, A. G., Groom, R. W., and Kurtz, R. D. (1993). Decomposition and modelling of the BC87 data set. *J. Geomag. Geoelectr.* **45**: 1127–1150.

Jones, F. W. and Pascoe, L. J. (1971). A general computer program to determine the perturbation of alternating electric currents in a two-dimensional model of a region of uniform conductivity with an embedded inhomogeneity. *Geophys. J. R. Astr. Soc.* **24**: 3–30.

(1972). The perturbation of alternating geomagnetic fields by three-dimensional conductivity inhomogeneities. *Geophys. J. R. Astr. Soc.* **27**: 479–485.

Jones, F. W. and Price, A. T. (1970). The perturbations of alternating geomagnetic fields by conductivity anomalies. *Geophys. J. R. Astr. Soc.* **20**: 317–334.

Jordan, T. H. (1978). Composition and development of the continental tectosphere. *Nature* **274**: 544–548.

Junge, A. (1990). A new telluric KCl probe using Filloux's Ag–AgCl electrode. *Pure and Applied Geophysics* **134**: 589–598.

Junge, A. (1994) Induzierte erdelektrische Felder–neue Beobachtungen in Norddeutschland und im Bramwald. Habilitation Thesis. Göttingen.

Karato, S. (1990). The role of hydrogen in the electrical conductivity of the upper mantle. *Nature* **347**: 272–273.

Karato, S. and Jung, H. (2003). Effects of pressure on high-temperature dislocation creep in olivine. *Phil. Mag.* **83**, 401–414.

Kariya, K. A. and Shankland, T. J. (1983). Electrical conductivity of dry lower crustal rocks. *Geophysics* **48**: 52–61.

Katsube, T. J. and Mareschal, M. (1993). Petrophysical model of deep electrical conductors: graphite lining as a source and its disconnection during uplift. *J. Geophys. Res.*, **98**: 8019–8030.

Keller, G. V. and Frischknecht, F. C. (1966). Electrical methods in geophysical prospecting. In *International Series of Monographs in Electromagnetic Waves*, **10**, eds. A.L. Cullen, V. A. Fock, and J. R. Wait. Oxford: Pergammon Press.

Kellet, R. L., Mareschal, M. and Kurtz, R. D. (1992). A model of lower crustal electrical anisotropy for the Pontiac Subprovince of the Canadian Shield. *Geophys. J. Int.* **111**: 141–150.

Kemmerle, K. (1977). On the influence of local anomalies of conductivity at the Earth's surface on magnetotelluric data. *Acta Geodaet., Geophys. et Montanist. Acad. Sci. Hung.* **12**: 177–181.

Key, K. and Constable, S. C. (2002). Broadband marine MT exploration of the East Pacific Rise at 9°50′N. *Geophys. Res. Lett.* **29**: (22), 2054 doi:10.1029/2002GL016035.

Keyser, M., Ritter, J. R. R. and Jordan, M. (2002). 3D shear-wave velocity structure of the Eifel plume, Germany. *Earth Planet. Sci. Lett.* **203**: 59–82.

Kohlstedt, D. L. and Mackwell, S. (1998). Diffusion of hydrogen and intrinsic point defects in olivine. *Z. Phys. Chem.* **207**: 147–162.

Koslovskaya, E. and Hjelt, S.-E. (2000). Modeling of elastic and electrical properties of solid–liquid rock system with fractal microstructure. *Phys. Chem. Earth (A)* **25**: 195–200.

Kozlovsky, Y. A. (1984). The world's deepest well. *Scientific American,* **251**: 106–112.

Kuckes, A. F. (1973). Relations between electrical conductivity of a mantle and fluctuating magnetic fields. *Geophys. J. R. Astr. Soc.* **32**: 119–131.

Kurtz, R. D., Craven, J. A., Niblett, E. R. and Stevens, R. A. (1993). The conductivity of the crust and mantle beneath the Kapuskasing uplift: electrical anisotropy in the upper mantle. *Geophys. J. Int.* **113**: 483–498.

Kuvshinov, A. V., Olsen, N., Avdeev, D. B. and Pankratov, O. V. (2002). Electromagnetic induction in the oceans and the anomalous behaviour of coastal C-responses for periods up to 20 days. *Geophys. Res. Lett.* **29**: (12), doi:10.1029/2002GL014409.

Lahiri, B. N. and Price, A. T. (1939). Electromagnetic induction in non-uniform conductors, and the determination of the conductivity of the Earth from terrestrial magnetic variations. *Phil. Trans. Roy. Soc. London (A)* **237**: 509–540.

Large, D. J., Christy, A. G. and Fallick, A. E. (1994). Poorly crystalline carbonaceous matter in high grade metasediments: implications for graphitisation and metamorphic fluid compositions. *Contrib. Mineral. Petrol.* **116**: 108–116.

Larsen, J. C. (1975). Low frequency (0.1–6.0 cpd) electromagnetic study of deep mantle electrical conductivity beneath the Hawaiian islands. *Geophys. J. R. Astr. Soc.* **43**: 17–46.

Larsen, J. C., Mackie, R. L., Manzella, A., Fiordelisi, A. and Rieven, S. (1996). Robust smooth magnetotelluric transfer functions. *Geophys. J. Int.* **124**: 801–819.

Lee, C. D., Vine, F. J. and Ross, R. G. (1983). Electrical conductivity models for the continental crust based on high grade metamorphic rocks. *Geophys. J. R. Astr. Soc.* **72**: 353–372.

Léger, A., Mathez, E. A., Duba, A., Pineau, F. and Ginsberg, S. (1996). Carbonaceous material in metamorphosed carbonate rocks from the Waits River Formation, NE Vermont, and its effect on electrical conductivity. *J. Geophys. Res.*, **101**: 22 203–22 214.

Leibecker, J., Gatzemeier, A., Hönig, M., Kuras, O. and Soyer, W. (2002). Evidence of electrical anisotropic structures in the lower crust and the upper mantle beneath the Rhenish Shield. *Earth Planet Sci. Lett.* **202**: 289–302.

Li, S., Unsworth, M., Booker, J. R., Wie, W., Tan, H. and Jones, A. G. (2003). Partial melt or aqueous fluid in the mid-crust of Southern Tibet? Constraints from INDEPTH magnetotelluric data. *Geophys. J. Int.* **153**: 289–304.

Lines, L. R. and Jones, F. W. (1973). The perturbation of alternating geomagnetic fields by three-dimensional island structures. *Geophys. J. R. Astr. Soc.* **32**: 133–154.

Lizzaralde, D., Chave, A., Hirth, G. and Schultz, A. (1995). Northeastern Pacific mantle conductivity profile from long-period magnetotelluric sounding using Hawaii-to-California submarine cable data. *J. Geophys. Res.* **100**: 17837–17854.

Mackie, R. L., Bennett, B. R. and Madden, T. R. (1988). Long-period MT measurements near the central California coast: a land-locked view of the conductivity structure under the Pacific ocean. *J. Geophys. Res.* **95**: 181–194.

Mackie, R. L. and Madden, T. R. (1993). Conjugate direction relaxation solutions for 3D magnetotelluric modelling. *Geophysics* **58**: 1052–1057.

Mackie, R. L., Madden, T. R. and Wannamaker, P. E. (1993). Three-dimensional magnetotelluric modeling using difference equations – Theory and comparisons to integral equation solutions. *Geophysics* **58**: 215–226.

Mackie, R. L., Rieven, S. and Rodi, W. (1997). *Users Manual and Software Documentation for Two-Dimensional Inversion of Magnetotelluric Data*. San Francisco: GSY-USA Inc.

Mackwell, S. J. and Kohlstedt, D. L. (1990). Diffusion of hydrogen in olivine: implications for water in the mantle. *J. Geophys. Res.* **95**: 5079–5088.

Madden, T. R. (1976). Random networks and mixing laws. *Geophysics*, **41**: 1104–1125.

Madden, T. and Nelson, P. (1964, reprinted 1986). A defence of Cagniard's magnetotelluric method. In *Society of Exploration Geophysicists, Geophysics Reprint Series, No. 5.*, ed. K. Vozoff.

Manoj, C. and Nagarajan, N. (2003). The application of artificial neural networks to magnetotelluric time-series analysis. *Geophys. J. Int.* **153**: 409–423.

Mareschal, M. (1986). Modelling of natural sources of magnetospheric origin in the interpretation of regional studies: a review. *Surv. Geophys.* **8**: 261–300.

Mareschal, M., Fyfe, W. S., Percival, J. and Chan, T. (1992). Grain-boundary graphite in Kapuskasing gneisses and implication for lower crustal conductivity. *Nature* **357**: 674–676.

Mareschal, M., Kellett, R. L., Kurtz, R. D., Ludden, J. N. and Bailey, R. C. (1995). Archean cratonic roots, mantle shear zones and deep electrical anisotropy. *Nature* **375**: 134–137.

Mareschal, M., Kurtz, R. D., Chouteau, M. and Chakridi, R. (1991). A magnetotelluric survey on Manitoulin Island and Bruce Peninsula along GLIMPCE seismic line J: black shales mask the Grenville Front. *Geophys. J. Int.* **105**: 173–183.

Masero, W., Fischer, G. and Schnegg, P.-A. (1997). Electrical conductivity and crustal deformation from magnetotelluric results in the region of the Araguainha impact, Brazil. *Phys. Earth Planet. Inter.* **101**: 271–289.

Mathez, E. A., Duba, A. G., Peach, C. L., Léger, A., Shankland, T. J. and Plafker, G. (1995). Electrical conductivity and carbon in metamorphic rocks of the Yukon-Tanana Terrane, Alaska. *J. Geophys. Res.*, **196**: 10187–10196.

Mathur, S. P. (1983). Deep crustal reflection results from the central Eromanga Basin, Australia. *Tectonophysics* **100**: 163–173.

Matsumoto, T., Honda, M., McDougall, I., Yatsevich, I. and O'Reilly, S. (1997). Plume-like neon in a metasomatic apatite from the Australian lithospheric mantle. *Nature* **388**: 162–164.

McKenzie, D. (1979). Finite deformation during fluid flow. *Geophys. J. R. Astr. Soc.* **58**: 689–715.

Meju, M. A. (1996). Joint inversion of TEM and distorted MT soundings: some effective practical considerations. *Geophysics* **61**: 56–65.

(2002). Geoelectromagnetic exploration for natural resources: models, case studies and challenges. *Surv. Geophys.* **23**: 133–206.

Fontes, S. L., Oliveira, M. F. B., Lima, J. P. R., Ulugergerli, E. U. and Carrasquilla, A. A. (1999). Regional aquifer mapping using combined VES-TEM-AMT / EMAP methods in the semiarid eastern margin of Parnaiba Basin, Brazil. *Geophysics* **64**: 337–356.

Merzer, A. M. and Klemperer, S. L. (1992). High electrical conductivity in a model lower crust with unconnected, conductive, seismically reflective layers. *Geophys. J. Int.* **108**: 895–905.

Morris, J. D., Leeman, W. P. and Tera, F. (1990). The subducted component in island arc lavas: constraints from Be isotopes and B–Be systematics. *Nature* **344**: 31–36.

Nelson, K. D. (1991). A unified view of craton evolution motivated by recent deep seismic reflection and refraction results. *Geophys. J. Int.* **105**: 25–35.

Nesbitt, B. E. (1993). Electrical resistivities of crustal fluids. *J. Geophys. Res.* **98**: 4301–4310.

Newton, R. C., Smith, J. V. and Windley, B. F. (1980). Carbonic metamorphism, granulites and crustal growth. *Nature* **288**: 45–50.

Nolasco, R., Tarits, P., Filloux, J. H. and Chave, A. D. (1998). Magnetotelluric imaging of the Society Island hotspot. *J. Geophys. Res.* **103**: 30287–30309.

Oettinger, G., Haak, V. and Larsen, J. C. (2001). Noise reduction in magnetotelluric time-series with a new signal-noise separation method and its application to a field experiment in the Saxonian Granulite Massif. *Geophys. J. Int.* **146**: 659–669.

Ogawa, Y. and Uchida, T. (1996). A two-dimensional magnetotelluric inversion assuming Gaussian static shift. *Geophys. J. Int.* **126**: 69–76.

Ogunade, S. O. (1995). Analysis of geomagnetic variations in south-western Nigeria. *Geophys. J. Int.* **121**: 162–172.

Olsen, N. (1998). The electrical conductivity of the mantle beneath Europe derived from C-responses from 3 to 720 hr. *Geophys. J. Int.* **133**: 298–308.
    (1999). Induction studies with satellite data. *Surv. Geophys.* **20**: 309–340.

Osipova, I. L., Hjelt, S.-E. and Vanyan, L. L. (1989). Source field problems in northern parts of the Baltic Shield. *Phys. Earth Planet. Inter.* **53**: 337–342.

Otnes, R. K. and Enochson, L. (1972). *Digital Time Series Analysis.* New York: John Wiley & Sons.

Padovani, E. and Carter, J. (1977). Aspects of the deep crustal evolution beneath south central New Mexico. In *The Earth's Crust: Its Nature and Physical Properties*, ed. J. G. Heacock, *Amer. Geophys. Union Monogr. Series*, **20**: 19–55.

Pádua, M. B., Padilha, A. L. and Vitorello, Í. (2002). Disturbances on magneto-telluric data due to DC electrified railway: a case study from southeastern Brazil. *Earth, Planets and Space* **54**: 591–596.

Park, S. K. (1985). Distortion of magnetotelluric sounding curves by three-dimensional structures. *Geophysics* **50**: 785–797.

Parker, E. N. (1958). Dynamics of the interplanetary gas and magnetic field. *Astrophys. J.* **128**: 664–676.

Parker, R. L. (1980). The inverse problem of electromagnetic induction: exist-ence and construction of solutions based on incomplete data. *J. Geophys. Res.* **85**: 4421–4428.

Parker, R. L. and Whaler, K. A. (1981). Numerical methods for establishing solutions to the inverse problem of electromagnetic induction. *J. Geophys. Res.* **86**: 9574–9584.

Parkinson, W. (1959). Directions of rapid geomagnetic variations. *Geophys. J. R. Astr. Soc.* **2**: 1–14.

Parkinson, W. (1971). An analysis of the geomagnetic diurnal variation during the IGY. *Gerlands Beitr. Geophys.* **80**: 199–232.

Parzen, E. (1961). Mathematical considerations in the estimation of spectra: comments on the discussion of Messers, Tukey and Goodman. *Technometrics* **3**: 167–190, 232–234.

Parzen, E. (1992). *Modern Probability Theory and its Applications.* New York: John Wiley & Sons.

Pek, J. and Verner, T. (1997). Finite-difference modelling of magnetotel-luric fields in two-dimensional anisotropic media. *Geophys. J. Int.* **128**: 505–521.

Petiau, G. and Dupis, A. (1980). Noise, temperature coefficient and long time stability of electrodes for telluric observations. *Geophys. Prosp.* **28**: 792–804.

Pous, J., Heise, W., Schnegg P.-A., Munoz, G., Marti, J. and Soriano, C. (2002). Magnetotelluric study of Las Cañadas caldera (Tenerife, Canary Islands): structural and hydrogeological implications. *Earth Planet. Sci. Lett.* **204**: 249–263.

Prácser, E. and Szarka, L. (1999). A correction to Bahr's 'phase deviation' method for tensor decomposition. *Earth, Planets and Space* **51**: 1019–1022.

Praus, O., Pecova, J., Petr, V, Babuska, V. and Plomerova, J. (1990). Magnetotelluric and seismological determination of the lithosphere–asthenosphere transition in Central Europe. *Phys. Earth Planet. Inter.* **60**: 212–228.

Presnall, D. C., Simmons, C.L. Porath, H. (1972). Changes in electriacal conductivitty of a syenthetic basalt during melting. *J. Geophys. Res.* **77**: 5665–5672.

Price, A. T. (1962). The theory of magnetotelluric fields when the source field is considered. *J. Geophys. Res.* **67**: 1907–1918.

Primdahl, F. (1979). The fluxgate magnetometer. *J. Phys. E: Sci. Instrum.* **12**: 241–253.

Raiche, A. P. (1974). An integral equation approach to three-dimensional modelling. *Geophys. J. R. Astr. Soc.* **36**: 363–376.

Ranganayaki, R. P. (1984). An interpretative analysis of magnetotelluric data. *Geophysics* **49**: 1730–1748.

Ranganayaki, R. P. and Madden, T. R. (1980). Generalized thin sheet analysis in magnetotellurics: an extension of Price's analysis. *Geophys. J. R. Astr. Soc.* **60**: 445–457.

Rangarajan, G. K. (1989). Indices of geomagnetic activity. In *Geomagnetism, Volume 3*, ed. J. A. Jacobs. London: Academic Press, pp. 323–384.

Rauen, A. and Laštovičková, M. (1995). Investigation of electrical anisotropy in the deep borehole KTB. *Surv. Geophys.* **16**: 37–46

Reddy, I. K., Rankin, D. and Phillips, R. J. (1977). Three-dimensional modelling in magnetotelluric and magnetic variational sounding. *Geophys. J. R. Astr. Soc.* **51**: 313–325.

Reynolds, J. M. (1997). *An Introduction to Applied and Environmental Geophysics*. New York: John Wiley & Sons.

Ribe, N. M. (1989). Seismic anisotropy and mantle flow. *J. Geophys. Res.* **94**: 4213–4223.

Ritter, P. and Banks, R. J. (1998). Separation of local and regional information in distorted GDS response functions by hypothetical event analysis. *Geophys. J. Int.* **135**: 923–942.

Ritter, J. R. R., Jordan, M., Christensen, U. R. and Achauer, U. (2001). A mantle plume below the Eifel volcanic fields, Germany. *Earth Planet. Sci. Lett.* **186**: 7–14.

Roberts, J. J., Duba, A. G., Mathez, E. A., Shankland, T. J. and Kinsler, R. (1999). Carbon-enhanced electrical conductivity during fracture of rocks. *J. Geophys. Res.* **104**: 737–747.

Roberts, J. J. and Tyburczy, J. A. (1991). Frequency-dependent electrical properties of polycrystalline olivine compacts. *J. Geophys. Res.* **96**: 16205–16222.

Roberts, J. J. and Tyburczy, J. A. (1999). Partial-melt electrical conductivity: influence of melt-composition. *J. Geophys. Res.* **104**: 7055–7065.

Ross, J. V. and Bustin, R. M. (1990). The role of strain energy in creep graphitisation of anthracite. *Nature* **343**: 58–60.

Rumble, D. R. and Hoering, T. C. (1986). Carbon isotope geochemistry of graphite vein deposits from New Hampshire, U.S.A. *Geochim. Cosmochim. Acta* **50**: 1239–1247.

Schilling, F. R., Partzsch, G. M., Brasse, H. and Schwarz, G. (1997). Partial melting below the magmatic arc in the central Andes deduced from geoelectromagnetic field experiments and laboratory data. *Phys. Earth Planet. Inter.* **103**: 17–31.

Schmeling, H. (1985). Numerical models on the influence of partial melt on elastic, anelastic and electrical properties of rocks. Part I: elasticity and anelasticity. *Phys. Earth Planet. Inter.* **41**: 105–110.

(1986). Numerical models on the influence of partial melt on elastic, anelastic and electrical properties of rocks. Part II: electrical conductivity. *Phys. Earth Planet. Inter.* **43**: 123–136.

Schmucker, U. (1970). Anomalies of geomagnetic variations in the Southwestern United States. *Bull. Scripps Inst. Ocean., University of California, 13*.

(1973). Regional induction studies: a review of methods and results. *Phys. Earth Planet. Inter.* **7**: 365–378.

(1978). Auswertungsverfahren Göttingen. In *Protokoll Kolloquium Elektromagnetische Tiefenforschung,* eds. V. Haak and J. Homilius. Free University Berlin, pp. 163–189.

(1987). Substitute conductors for electromagnetic response estimates. *Pure and Appl. Geophys.* **125**: 341–367.

(1995). Electromagnetic induction in thin sheets: integral equations and model studies in two dimensions. *Geophys. J. Int.* **121**: 173–190.

Schultz, A., Kurtz, R. D., Chave, A. D. and Jones, A. G. (1993). Conductivity discontinuities in the upper mantle beneath a stable craton. *Geophys. Res. Lett.* **20**: 2941–2944.

Schuster, A. (1889). The diurnal variation of terrestrial magnetism. *Phil. Trans. Roy. Soc. London* **A210**: 467–518.

Sénéchal, R., Rondenay, G. S., Mareschal, M., Guilbert, J. and Poupinet, G. (1996). Seismic and electrical anisotropies in the lithosphere across the Grenville Front, Canada. *Geophys. Res. Lett.* **23**: 2255–2258.

Shankland, T. J. and Ander, M. E. (1983). Electrical conductivity, temperatures, and fluids in the lower crust. *J. Geophys. Res.* **88**: 9475–9484.

Shankland, T. J., Duba, A., Mathez, E. A. and Peach, C. L. (1997). Increase of electrical conductivity with pressure as an indication of conduction through a solid phase in mid-crustal rocks. *J. Geophys. Res.* **102**: 14741–14750.

Shankland, T. J., Peyronneau, J. and Poirier, J.-P. (1993). Electrical conductivity of the Earth's lower mantle. *Nature* **366**: 453–455.

Shankland T. J. and Waff, H. S. (1977). Partial melting and electrical conductivity anomalies in the upper mantle. *J. Geophys. Res.* **82**: 5409–5417.

Siegesmund, S., Vollbrecht, A. and Nover, G. (1991). Anisotropy of compressional wave velocities, complex electrical resistivity and magnetic susceptibility of mylonites from the deeper crust and their relation to the rock fabric. *Earth Planet. Sci. Lett.* **105**: 247–259.

Siemon, B. (1997). An interpretation technique for superimposed induction anomalies. *Geophys. J. Int.* **130**: 73–88.

Simpson, F. (1999). Stress and seismicity in the lower continental crust: a challenge to simple ductility and implications for electrical conductivity mechanisms. *Surveys in Geophysics* **20**: 201–227.

(2000). A three-dimensional electromagnetic model of the southern Kenya Rift: departure from two-dimensionality as a possible consequence of a rotating stress field. *J. Geophys. Res.* **105**: 19321–19334.

(2001a). Fluid trapping at the brittle–ductile transition re-examined. *Geofluids* **1**: 123–136.

(2001b). Resistance to mantle flow inferred from the electromagnetic strike of the Australian upper mantle. *Nature* **412**: 632–635.

(2002a). Intensity and direction of lattice-preferred orientation of olivine: are electrical and seismic anisotropies of the Australian mantle reconcilable? *Earth Planet Sci. Lett.* **203**: 535–547.

(2002b). A comparison of electromagnetic distortion and resolution of upper mantle conductivities beneath continental Europe and the Mediterranean using islands as windows. *Phys. Earth Planet. Inter.* **129**: 117–130.

Simpson, F. and Warner, M. (1998). Coincident magnetotelluric, *P*-wave and *S*-wave images of the deep continental crust beneath the Weardale granite, NE England: seismic layering, low conductance and implications against the fluids paradigm. *Geophys. J. Int.* **133**: 419–434.

Sims, W. E., Bostick, F. X. Jr. and Smith, H. W. (1971). The estimation of magnetotelluric impedance tensor elements from measured data. *Geophysics* **36**: 938–942.

Smith, T. and Booker, J. (1988). Magnetotelluric inversion for minimum structure. *Geophysics* **53**: 1565–1576.

(1991). Rapid inversion of two- and three-dimensional magnetotelluric data. *J. Geophys. Res.*, **96**: 3905–3922.

Soyer, W. and Brasse, H. (2001). A magneto-variation array study in the central Andes of N Chile and SW Bolivia. *Geophys. Res. Lett.* **28**: 3023–3026.

Spitzer, K. (1993). Observations of geomagnetic pulsations and variations with a new borehole magnetometer down to depths of 3000 m. *Geophys. J. Int.* **115**: 839–848.

(1995). A 3-D finite difference algorithm for dc resistivity modelling using conjugate gradient methods. *Geophys. J. Int.* **123**: 903–914.

Srivastava, S. P. (1965). Methods of interpretation of magnetotelluric data when the source field is considered. *J. Geophys. Res.* **70**: 945–954.

Stacey, F. D. (1992). *Physics of the Earth*. Brisbane: Brookfield Press.

Stalder, R. and Skogby, H. (2003). Hydrogen diffusion in natural and synthetic orthopyroxene. *Phys. Chem. Minerals* **30**: 12–19.

Stanley, W. D. (1989). Comparison of geoelectrical/tectonic models for suture zones in the western U.S.A. and eastern Europe: are black shales a possible source of high conductivities? *Phys. Earth Planet. Inter.* **53**: 228–238.

Stauffer, D. and Aharony, A. (1992). *Introduction to Percolation Theory*, 2nd edn. London: Taylor and Francis.

Sternberg, B. K., Washburne, J. C. and Pellerin, L. (1988). Correction for the static shift in magnetotellurics using transient electromagnetic soundings. *Geophysics* **53**: 1459–1468.

Stesky, R. M. and Brace, W. F. (1973). Electrical conductivity of serpentinized rocks to 6 kilobars. *J. Geophys. Res.* **78**: 7614–7621.

Stoerzel, A. (1996). Estimation of geomagnetic transfer functions from non-uniform magnetic fields induced by the equatorial electrojet: a method to determine static shifts in magnetotelluric data. *J. Geophys. Res.* **101**: 917–927.

Strack, K. M. (1992). *Exploration with Deep Transient Electromagnetics.* Amsterdam: Elsevier.

Swift, C. M. (1967). A magnetotelluric investigation of an electrical conductivity anomaly in the South Western United States. Ph.D. Thesis, M.I.T., Cambridge, Mass.

(1986). A magnetotelluric investigation of an electrical conductivity anomaly in the South Western United States. In *Magnetotelluric Methods*, ed. K. Vozoff. Tulsa: Society of Exploration Geophysicists, pp. 156–166.

Tikhonov, A. N. (1950). The determination of the electrical properties of deep layers of the Earth's crust. *Dokl. Acad. Nauk. SSR* **73**: 295–297 (in Russian).

(1986). On determining electrical characteristics of the deep layers of the Earth's crust. In *Magnetotelluric Methods*, ed. K. Vozoff. Tulsa: Society of Exploration Geophysicists, pp. 2–3.

Ting, S. C. and Hohmann, G. W. (1981). Integral equation modeling of three-dimensional magnetotelluric response. *Geophysics* **46**: 182–197.

Tipler, P. A. (1991). *Physics for Scientists and Engineers.* New York: Worth Publishers.

Torres-Verdin, C. and Bostick, F. X. (1992). Principles of spatial surface electric field filtering in magnetotellurics: electromagnetic array profiling (EMAP). *Geophysics* **57**: 603–622.

Touret, J. (1986). Fluid inclusions in rocks from the lower continental crust. In *The Nature of the Lower Continental Crust,* eds. J. B. Dawson, D. A. Carswell, J. Hall and K. H. Wedepohl. *Geological Society Special Publication* **24**: 161–172. Oxford: Blackwell Scientific Publications.

Tyburczy, J. A. and Waff, H. S. (1983). Electrical conductivity of molten basalt and andersite to 25 kilobars pressure: geophysical significance and implications for charge transport and melt structure. *J. Geophys. Res.* **88**: 2413–2430.

Valdivia, J. A., Sharma, A. S. and Papadopoulos, K. (1996). Prediction of magnetic storms by nonlinear models. *Geophys. Res. Lett.* **23**: 2899–2902.

Vanyan, L. L. and Gliko, A. O. (1999). Seismic and electromagnetic evidence of dehydration as a free water source in the reactivated crust. *Geophys. J. Int.* **137**: 159–162.

Vasseur, G. and Weidelt, P. (1977). Bimodal electromagnetic induction in non-uniform thin sheets with an application to the northern Pyrenean induction anomaly. *Geophys. J. R. Astr. Soc.* **51**: 669–690.

Vozoff, K. (1972). The magnetotelluric method in the exploration of sedimentary basins. *Geophysics* **37**: 98–141.

Waff, H. S. (1974). Theoretical considerations on electrical conductivity in a partially molten mantle and implications for geothermometry. *J. Geophys. Res.* **79**: 4003–4010.

Wait, J. R. (1954). On the relation between telluric currents and the Earth's magnetic field. *Geophysics* **19**: 281–289.

Wang, L.J and Lilley, F. E. M. (1999). Inversion of magnetometer array data by thin-sheet modelling. *Geophys. J. Int.* **137**: 128–138.

Wang, L., Zhang, Y. and Essene, E. (1996). Diffusion of the hydrous component in pyrope. *Amer. Mineral.* **81**: 706–718.

Wannamaker, P. E., Hohmann G. W. and Ward, S. H. (1984a). Magnetotelluric responses of three-dimensional bodies in layered Earths. *Geophysics* **49**: 1517–1533.

Wannamaker, P. E., Hohmann G. W. and San Filipo W. A. (1984b). Electromagnetic modelling of three-dimensional bodies in layered Earths using integral equations. *Geophysics* **48**: 1402–1405.

Wannamaker, P. E., Stodt, J. A. and Rijo, L. (1986). A stable finite element solution for two-dimensional magnetotelluric modelling. *Geophys. J. R. Astr. Soc.* **88**: 277–296.

Wannamaker, P. E. (2000). Comment on 'The petrological case for a dry lower crust' by Bruce W. D. Yardley and John W. Valley. *J. Geophys. Res.* **105**: 6057–6064.

Weaver, J. T. (1994). *Mathematical Methods for Geo-Electromagnetic Induction.* Taunton, Somerset, UK: Research Studies Press Ltd.

Wei, W., Unsworth, M., Jones, A. G. *et al.* (2001). Detection of widespread fluids in the Tibetan crust by magnetotelluric studies. *Science* **292**: 716–718.

Weidelt, P. (1972). The inverse problem of geomagnetic induction. *Z. Geophys.* **38**: 257–289.

(1975). Electromagnetic induction in three-dimensional structures. *J. Geophys. Res.* **41**: 85–109.

(1985). Construction of conductance bounds from magnetotelluric impedances. *J. Geophys.* **57**: 191–206.

Wiese, H. (1962). Geomagnetische tiefensondierung. Teil II: Die Streichrichtung der Untergrundstrukturen des elektrischen Widerstandes, erschlossen aus geomagnetischen variationen. *Geofis. Pura et Appl.* **52**: 83–103.

Winch, D. E. (1981). Spherical harmonic analysis of geomagnetic tides, 1964–1965. *Phil. Trans. Roy. Soc. Lond.* **A303**: 1–104.

Woods, S. C., Mackwell, S. and Dyar, D. (2000). Hydrogen in diopside: diffusion profiles. *Amer. Mineral.* **85**: 480–487.

Xu, Y., Poe, B. T., Shankland, T. J. and Rubie, D. C. (1998). Electrical conductivity of olivine, wadsleyite, and ringwoodite under upper-mantle conditions. *Science* **280**: 1415–1418.

Xu, Y. and Shankland, T. J. (1999). Electrical conductivity of orthopyroxene and its high pressure phases. *Geophys. Res. Lett.* **26**: 2645–2648.

Xu, Y., Shankland, T. J. and Poe, B. T. (2000). Laboratory-based electrical conductivity in the Earth's mantle. *J. Geophys. Res.* **105**: 27865–27875.

Yardley, B. W. D. (1986). Is there water in the deep continental crust? *Nature* **323**: 111.

Yardley, B. W. D. and Valley, J. W. (1997). The petrological case for a dry lower crust. *J. Geophys. Res.* **102**: 12173–12185.

Yardley, B. W. D. and Valley, J. W. (2000). Reply to Wannamaker (2000) "Comment on 'The petrological case for a dry lower crust'" *J. Geophys. Res.* **105**: 6065–6068.

Zhdanov, M. S., Varentsov, I. M., Weaver, J. T., Golubev, N. G. and Krylov, V. A. (1997). Methods for modelling electromagnetic fields. Results from COMMEMI – the international project on the comparison of modelling methods for electromagnetic induction. *J. Appl. Geophys.* **37**: 133–271.

Zhang, P., Pedersen, L. B., Mareschal, M. and Chouteau, M. (1993). Channelling contribution to tipper vectors: a magnetic equivalent to electrical distortion. *Geophys. J. Int.* **113**: 693–700.

Zonge, K. L. and Hughes, L. H. (1991). Controlled-source audio-frequency magnetotellurics. In *Electromagnetic Methods in Applied Geophysics. Volume 2: Applications, Part B.*, ed. M. C. Nabighian. Tulsa: Society of Exploration Geophysicists, pp. 713–809.

Index

Bold entries are defined in the glossary. *Italic* page numbers refer to a figure.

United States
ers

Printed in th
By Bookma